Analysis of Oriented Texture

with Applications to the
Detection of Architectural Distortion in Mammograms

Analysis of Oriented Texture - with Applications to the Detection of Architectural Distortion in Mammograms
Fábio J. Ayres, Rangaraj M. Rangayyan, and J. E. Leo Desautels

ISBN: 978-3-031-00519-0   paperback
ISBN: 978-3-031-01647-9   ebook

DOI 10.1007/978-3-031-01647-9

A Publication in the Springer series
SYNTHESIS LECTURES ON BIOMEDICAL ENGINEERING

Lecture #38
Series Editor: John D. Enderle, *University of Connecticut*
Series ISSN
Synthesis Lectures on Biomedical Engineering
Print 1930-0328   Electronic 1930-0336

Synthesis Lectures on Biomedical Engineering

Editor
John D. Enderle, *University of Connecticut*

Lectures in Biomedical Engineering will be comprised of 75- to 150-page publications on advanced and state-of-the-art topics that spans the field of biomedical engineering, from the atom and molecule to large diagnostic equipment. Each lecture covers, for that topic, the fundamental principles in a unified manner, develops underlying concepts needed for sequential material, and progresses to more advanced topics. Computer software and multimedia, when appropriate and available, is included for simulation, computation, visualization and design. The authors selected to write the lectures are leading experts on the subject who have extensive background in theory, application and design.

The series is designed to meet the demands of the 21st century technology and the rapid advancements in the all-encompassing field of biomedical engineering that includes biochemical, biomaterials, biomechanics, bioinstrumentation, physiological modeling, biosignal processing, bioinformatics, biocomplexity, medical and molecular imaging, rehabilitation engineering, biomimetic nano-electrokinetics, biosensors, biotechnology, clinical engineering, biomedical devices, drug discovery and delivery systems, tissue engineering, proteomics, functional genomics, molecular and cellular engineering.

Analysis of Oriented Texture

with Applications to the

Detection of architectural Distortion in Mammograms

Fábio J. Ayres, Rangaraj M. Rangayyan, and J. E. Leo Desautels
University of Calgary

SYNTHESIS LECTURES ON BIOMEDICAL ENGINEERING #38

ABSTRACT

The presence of oriented features in images often conveys important information about the scene or the objects contained; the analysis of oriented patterns is an important task in the general framework of image understanding. As in many other applications of computer vision, the general framework for the understanding of oriented features in images can be divided into low- and high-level analysis. In the context of the study of oriented features, low-level analysis includes the detection of oriented features in images; a measure of the local magnitude and orientation of oriented features over the entire region of analysis in the image is called the orientation field. High-level analysis relates to the discovery of patterns in the orientation field, usually by associating the structure perceived in the orientation field with a geometrical model. This book presents an analysis of several important methods for the detection of oriented features in images, and a discussion of the phase portrait method for high-level analysis of orientation fields. In order to illustrate the concepts developed throughout the book, an application is presented of the phase portrait method to computer-aided detection of architectural distortion in mammograms.

KEYWORDS

oriented texture, architectural distortion, mammography, breast cancer, Gabor filters, phase portraits, line detectors, optimization techniques, computer-aided diagnosis

Contents

Preface

The presence of oriented features in images often conveys important information about the scene or the objects contained; the analysis of oriented patterns is an important task in the general framework of image understanding. As in many other applications of computer vision, the general framework for the understanding of oriented features in images can be divided into low- and high-level analysis. In the context of the study of oriented features, low-level analysis includes the detection of oriented features in images; a measure of the local magnitude and orientation of oriented features over the entire region of analysis in the image is called the orientation field. High-level analysis relates to the discovery of patterns in the orientation field, usually by associating the structure perceived in the orientation field with a geometrical model. This book presents an analysis of several important methods for the detection of oriented features in images, and a discussion of the phase portrait method for high-level analysis of orientation fields. In order to illustrate the concepts developed throughout the book, an application is presented of the phase portrait method to computer-aided detection of architectural distortion in mammograms.

The methods described in this book are mathematical in nature. It is assumed that the reader is proficient in advanced mathematics and familiar with basic notions of data, signal, and image processing. The methods of image analysis and modeling are suitable for inclusion in advanced courses in electrical engineering, computer engineering, mathematics, physics, computer science, biomedical engineering, and bioinformatics. The book is illustrated with several images and examples of application to facilitate efficient comprehension of the notions and methods presented.

We wish our readers success in their studies and research.

Fábio J. Ayres, Rangaraj M. Rangayyan, and J. E. Leo Desautels
November 2010

CHAPTER 1

Detection of Oriented Features in Images

1.1 ORIENTED FEATURES IN IMAGES

The presence of oriented features in images often conveys important information about the scene or the objects contained; the analysis of oriented patterns is an important task in the general framework of image understanding. Some examples of oriented pattern analysis are listed below:

- The degree of organization of individual fibril strands in paper and textiles is related to the strength of the material [1].

- Images obtained by remote sensing may present piecewise linear features related to rivers, ridges, troughs, and man-made structures such as roads, buildings, farmland, and urban spaces. The identification of natural directional patterns in satellite images is useful in the study of the underlying geological processes that formed the observed terrain [2, 3].

- Seismic images have linear elements (such as "ground roll") that must be eliminated in order to improve the accuracy of the interpretation of such images [4, 5, 6].

- Fingerprints are formed by the arrangement of linear features. Fingerprint analysis, recognition, and archival are capabilities essential to forensic investigators [7].

- Several biomedical images present oriented features, such as images of collagen fibers in ligaments [8, 9]; vascular networks in ligaments, lungs, and the heart [10]; bronchial trees in lungs [11, 12]; fibroglandular tissue, the pectoral muscle, ligaments, and ducts in the breast [13, 14, 15, 16, 30, 63, 143] (see Appendix A); and vascular networks in the retina [17].

- Wood-grain images may be analyzed to detect the presence of defects that might affect the strength of the material [18].

- The histogram of edge orientations has been used to detect the presence of a human being in an image [19].

- Printed circuit boards contain several linear elements that must be properly identified prior to the detection of defects [20].

- The detection of pen strokes is important in the context of handwriting recognition [21].

- The counting of asbestos fibers in a known volume of air is an important measure to monitor. Asbestos is widely recognized as a health hazard, despite its useful chemical and physical properties. Asbestos fibers appear in microscopy images as line segments, and algorithms have been developed to detect and count such fibers [22].

- Insects growing in bulk wheat grain may appear in inspection images as line segments. Computer vision systems have been developed for the automated inspection of cereal grains [23], which is desirable to monitor the quality of stocked grains.

- Some authors [24, 25] have investigated the description of natural images using PCA. It has been observed that the principal components of natural images have the characteristics of bandpass filters, and that the filters can be organized in groups where each group is a steerable basis of oriented filters [25].

Several methods have been developed for the analysis of directionally oriented texture and line detection, including the following:

- Fourier-transform-based methods [4, 5, 6, 9];

- Space-domain linear filters, such as steerable filters [26], Gabor filters [14, 27, 28, 29, 30], and infinite-impulse-response recursive fan filters [31]; and

- Nonlinear filters [32, 33].

The detection of oriented features is influenced by the characteristic width of the features under investigation and the presence of noise. Individual oriented features are associated with a particular spatial width, such as the width of a spicule in a mammographic image, or that of a road seen in an aerial or satellite image. Hence, the ability of a technique to detect oriented features depends on the proper calibration of a scale parameter that regulates the intrinsic width of the oriented feature detector, according to the width of the oriented features of interest. Noise is another factor that may impair the performance of an oriented feature detector because noise can mask the presence of oriented structures and may also form or lead to the detection of spurious oriented structures in the image.

In this chapter, criteria are developed to design and analyze the detection performance and orientation accuracy of five selected oriented feature detectors [34, 35]: the Gaussian second-derivative steerable filter [26], the quadrature-pair Gaussian second-derivative steerable filter [26], the real Gabor filter [27, 28], the complex Gabor filter [27, 28], and the line operator of Dixon and Taylor [22]. The detection performance is defined in terms of the ability of each filter to detect linear structures in the presence of noise and imprecision in the specification of scale. The orientation accuracy is derived in terms of the cumulative angle error for the pixels belonging to the oriented features in a test pattern.

1.2   DIRECTIONALLY SENSITIVE FILTERS

1.2.1   STEERABLE FILTERS

The concept of steerable filters was presented by Freeman and Adelson [26]. Steerable filters are a class of filters with the "steering property": the impulse response of a filter, rotated at an arbitrary angle, can be synthesized from a linear combination of rotated versions of the same filter's impulse response, for a set of pre-specified rotation parameters. Let $f(x, y)$ be the filter's impulse response, and $f_\theta(x, y)$ be the rotated impulse response at an arbitrary angle θ. The steering property can be written as

$$f_\theta(x, y) = \sum_{j=1}^{M} k_j(\theta) f_{\theta_j}(x, y),$$

where θ_j is the jth basis angle, $j \in 1, 2, \cdots, M$, and $k_j(\theta)$ is the jth interpolating function.

Gaussian-derivative-based steerable filters:

The partial derivatives of the Gaussian function provide a family of *steerable filters* [26]. In particular, the second derivative of the Gaussian has been used as a detector of linear structures in mammograms [15]. In the present analysis, the second derivative of the Gaussian is implemented in a form similar to that of the filter described in the work of Freeman and Adelson [26]:

$$s(x, y) = \frac{-0.9213}{\sigma_s} \left(2 \left(\frac{x}{\sigma_s} \right)^2 - 1 \right) \exp\left(-\frac{x^2 + y^2}{\sigma_s^2} \right),$$

where σ_s determines the scale of the filter. The integral of $|s(x, y)|^2$ over the $\mathbb{R}^2$ plane is normalized to unity. Figure 1.1 illustrates the steerable filter, for $\sigma_s = 1$.

This function requires three basis angles to implement the steering property: $\theta_1^s = 0$, $\theta_2^s = \pi/3$, and $\theta_3^s = 2\pi/3$. Then, the steering property can be written as

$$s_\theta(x, y) = \sum_{j=1}^{3} k_j^s(\theta) s_{\theta_j^s}(x, y),$$

where $s_\theta(x, y)$ denotes $s(x, y)$ rotated by the angle θ, and the interpolating functions $k_j^s(\theta)$ are given by

$$k_j^s(\theta) = \frac{1}{3} \left[1 + 2\cos(2(\theta - \theta_j^s)) \right].$$

Let $I(x, y)$ be the given image, and $W_\theta^s(x, y)$ be the result of filtering $I(x, y)$ with the filter $s_\theta(x, y)$, for a given value of θ. In the present work, frequency-domain filtering is employed: the

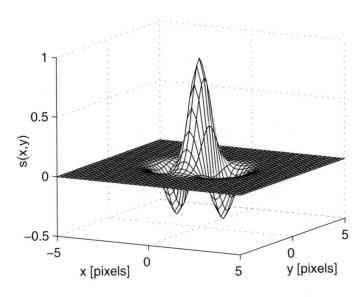

Figure 1.1: Basic steerable filter kernel, with $\sigma_s = 1$. Reproduced with permission from F. J. Ayres and R. M. Rangayyan. "Design and performance analysis of oriented feature detectors". Journal of Electronic Imaging, 16(2):12 pages, April 2007. article number 023007. © SPIE

Fourier transform of $I(x, y)$ is multiplied by the analytical expression for the Fourier transform of $s_\theta(x, y)$, and the inverse Fourier transform is applied to the resulting image in order to obtain $W_\theta^s(x, y)$.

The magnitude function $W_\theta^s(x, y)$ is periodic in θ; hence, it can be decomposed into a Fourier series. Let $C_1^s(x, y)$ and $C_2^s(x, y)$ be the cosine and sine coefficients, respectively, of the second harmonic of $W_\theta^s(x, y)$. Then, the local orientation $\phi^s(x, y)$ of $I(x, y)$ is defined as $\phi^s(x, y) = \theta_{opt}^s(x, y) + \pi/2$, where

$$\theta_{opt}^s(x, y) = \frac{1}{2} \arctan\left(\frac{C_2^s}{C_1^s}\right) = \frac{1}{2} \arctan\left(\frac{\sqrt{3}(W_{\pi/3}^s - W_{2\pi/3}^s)}{W_{\pi/3}^s + W_{2\pi/3}^s - 2W_0^s}\right).$$

The coordinates (x, y) have been dropped for clarity. The magnitude component of the orientation field is given by $E^s(x, y) = W_{\theta_{opt}}^s(x, y)$. The amplitude $E^s(x, y)$ and the local orientation $\phi^s(x, y)$ compose the orientation field extracted with the steerable Gaussian second-derivative filter. This procedure enhances oriented features of positive contrast only: the features of interest are required to be brighter than their surroundings.

Quadrature-pair Gaussian-derivative-based steerable filters: The Gaussian second-derivative steerable filter forms a quadrature pair with its Hilbert transform. Freeman and Adelson [26] present an approximation $s^h(x, y)$ of the Hilbert-transformed filter composed of a polynomial times a Gaussian function in (x, y). With the inclusion of the scale factor σ_s, the Hilbert-transformed steerable kernel analyzed in this chapter is designed as follows [26]:

$$s^h(x, y) = \frac{1}{\sigma_s}\left(-2.205\left(\frac{x}{\sigma_s}\right) + 0.9780\left(\frac{x}{\sigma_s}\right)^3\right)\exp\left(-\frac{x^2 + y^2}{2\sigma_s^2}\right).$$

The constants in the expression above are required to normalize the integral of $|s^h(x, y)|^2$ over the $\mathbb{R}^2$ plane to unity. Figure 1.2 illustrates the Hilbert-transformed steerable filter kernel, for $\sigma_s = 1$.

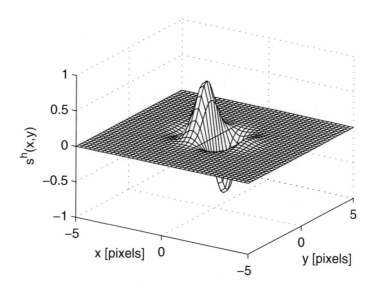

Figure 1.2: Basic Hilbert-transformed steerable filter kernel, with $\sigma_s = 1$. Reproduced with permission from F. J. Ayres and R. M. Rangayyan. "Design and performance analysis of oriented feature detectors". Journal of Electronic Imaging, 16(2):12 pages, April 2007. article number 023007. © SPIE

This function requires four basis angles to steer: $\theta_1^h = 0, \theta_2^h = \pi/4, \theta_3^h = \pi/2$, and $\theta_4^h = 3\pi/4$. The corresponding interpolation functions are given by

$$k_j^h(\theta) = \frac{1}{2}\left[\cos(\theta - \theta_j^h) + \cos(3(\theta - \theta_j^h))\right].$$

Analogous to the steerable Gaussian second-derivative filter, define $W_\theta^h(x, y)$ as the output of filtering the given image $I(x, y)$ with the filter $s_\theta^h(x, y)$. The filtering operation is performed in the frequency domain. The combined energy of the quadrature pair is given by $|W_\theta^{sh}(x, y)|^2 =$

$|W_\theta^s(x, y)|^2 + |W_\theta^h(x, y)|^2$, which can be decomposed in a Fourier series of only even harmonic frequencies in θ, for a given (x, y). (Only even harmonics appear in the Fourier series as a consequence of the squaring operation.) Let C_1^{sh} and C_2^{sh} be the cosine and sine coefficients, respectively, of the second harmonic of $|W_\theta^{sh}(x, y)|^2$. Then, the local orientation $\phi^{sh}(x, y)$ of $I(x, y)$ is defined as $\phi^{sh}(x, y) = \theta_{opt}^{sh}(x, y) + \pi/2$, where

$$\theta_{opt}^{sh}(x, y) = \frac{1}{2} \arctan\left(\frac{C_2^{sh}}{C_1^{sh}}\right)$$

and

$$C_1^{sh} = \frac{4}{9}(W_0^s)^2 + \frac{2}{9} W_0^s W_{\pi/3}^s + \frac{2}{9} W_0^s W_{2\pi/3}^s - \frac{2}{9}(W_{\pi/3}^s)^2$$

$$- \frac{4}{9} W_{\pi/3}^s W_{2\pi/3}^s - \frac{2}{9}(W_{2\pi/3}^s)^2 + \frac{3}{8}(W_0^h)^2$$

$$+ \frac{1}{8}\sqrt{2} W_0^h W_{\pi/4}^h - \frac{1}{8}\sqrt{2} W_0^h W_{3\pi/4}^h - \frac{1}{8}\sqrt{2} W_{\pi/4}^h W_{\pi/2}^h$$

$$+ \frac{1}{4} W_{\pi/4}^h W_{3\pi/4}^h - \frac{3}{8}(W_{\pi/2}^h)^2 - \frac{1}{8}\sqrt{2} W_{\pi/2}^h W_{3\pi/4}^h$$

$$C_2^{sh} = \frac{2}{9}\sqrt{3} W_0^s W_{\pi/3}^s - \frac{2}{9}\sqrt{3} W_0^s W_{2\pi/3}^s + \frac{2}{9}\sqrt{3}(W_{\pi/3}^s)^2$$

$$- \frac{2}{9}\sqrt{3}(W_{2\pi/3}^s)^2 + \frac{1}{8}\sqrt{2} W_0^h W_{\pi/4}^h - \frac{1}{4} W_0^h W_{\pi/2}^h$$

$$+ \frac{1}{8}\sqrt{2} W_0^h W_{3\pi/4}^h + \frac{3}{8}(W_{\pi/4}^h)^2 + \frac{1}{8}\sqrt{2} W_{\pi/4}^h W_{\pi/2}^h$$

$$- \frac{1}{8}\sqrt{2} W_{\pi/2}^h W_{3\pi/4}^h - \frac{3}{8}(W_{3\pi/4}^h)^2 .$$

The coordinates (x, y) have been dropped for clarity. The magnitude of the output of the quadrature-pair filter is given by $E^{sh}(x, y) = |W_{\theta_{opt}}^{sh}(x, y)|$. The magnitude $E^{sh}(x, y)$ and the local orientation $\phi^{sh}(x, y)$ compose the orientation field extracted with the steerable quadrature-pair Gaussian second-derivative filter. This filter detects oriented features of positive and negative contrast since the energy of the filter is a positive quantity in both cases.

The Gaussian second-derivative filter will be henceforth referred to as the steerable filter only; the quadrature-pair Gaussian second-derivative filter will be referred to as the quadrature steerable filter.

1.2.2   GABOR FILTERS

Gabor filters are a category of filters obtained from the modulation of a sinusoidal function (real or complex) by a Gaussian envelope [27]. The Gabor filter represents the best compromise between spatial localization and frequency localization, as measured by the product between the spatial extent and the frequency bandwidth of the filter [27, 28]. In image processing applications, Gabor filters may be used as oriented feature detectors [14, 29, 30, 36].

The real Gabor filter:

The real Gabor filter kernel oriented at the angle $\theta = -\pi/2$ is given by

$$g^r(x, y) = \frac{1}{2\pi \sigma_x \sigma_y} \exp\left[-\frac{1}{2}\left(\frac{x^2}{\sigma_x^2} + \frac{y^2}{\sigma_y^2}\right)\right] \cos(2\pi f x). \tag{1.1}$$

Kernels at other angles can be obtained by rotating this kernel over the range $[-\pi/2, \pi/2]$ by using the coordinate transformation

$$\begin{bmatrix} x' \\ y' \end{bmatrix} = \begin{bmatrix} \cos\alpha & \sin\alpha \\ -\sin\alpha & \cos\alpha \end{bmatrix} \begin{bmatrix} x \\ y \end{bmatrix}$$

where (x', y') is the set of coordinates rotated by the angle α. In this study, the parameters in Equation 1.1, namely σ_x, σ_y, and f, are derived from design rules as follows [14]:

- Let τ be the full-width at half-maximum of the Gaussian term in Equation 1.1 along the x axis. Then, $\sigma_x = \tau/(2\sqrt{2\ln 2}) = \tau/2.35$.

- The cosine term has a period of τ; therefore, $f = 1/\tau$.

- The value of σ_y is defined as $\sigma_y = l\,\sigma_x$, where l determines the elongation of the Gabor filter in the y direction, as compared to the extent of the filter in the x direction. In this work, we use $l = 8$.

The parameter τ controls the scale of the filter. Figure 1.3 illustrates the real Gabor filter kernel, for $\tau = 4$ pixels and $l = 8$.

Let $\phi^r(x, y)$ be the angle of the oriented feature at (x, y), and $g_k^r(x, y), k = 0, 1, \cdots, (K-1)$, be the real Gabor filter oriented at $\alpha_k = -\pi/2 + \pi k/K$, where K is the number of equally spaced filters over the angular range $[-\pi/2, \pi/2]$; in this analysis, let $K = 18$ with synthetic images and $K = 180$ with mammograms. The functions $g_k^r(x, y), k = 0, 1, \cdots, (K-1)$ form a bank of real Gabor filters, from which the orientation field can be extracted.

Let $I(x, y)$ be the image being processed, and $W_k^r(x, y)$ represent the Gabor-filtered images. The filtering operation is implemented in the frequency domain, analogous to the steerable filter case. Then, the orientation field of $I(x, y)$ produced by the bank of real Gabor filters is given by the angle

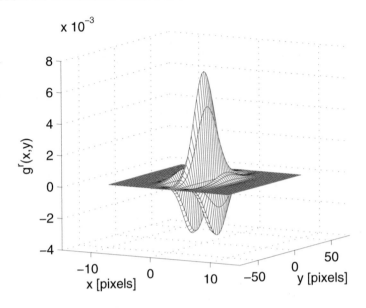

Figure 1.3: Basic real Gabor filter kernel, with $\tau = 4$ and $l = 8$. Reproduced with permission from F. J. Ayres and R. M. Rangayyan. "Design and performance analysis of oriented feature detectors". Journal of Electronic Imaging, 16(2):12 pages, April 2007. article number 023007. © SPIE

$$\phi^r(x, y) = \alpha_{k_{\max}} \quad \text{where} \quad k_{\max} = \arg\{\max_k[W_k^r(x, y)]\} \,,$$

and by the amplitude of the output of the real Gabor filter at the optimal orientation $E^r(x, y) = W_{k_{\max}}^r(x, y)$. This filter is sensitive to oriented features of positive contrast only.

The complex Gabor filter:

The complex Gabor filter kernel oriented at the angle $\theta = -\pi/2$ is given by

$$
\begin{aligned}
g^c(x, y) &= \frac{1}{2\pi \sigma_x \sigma_y} \exp\left[-\frac{1}{2}\left(\frac{x^2}{\sigma_x^2} + \frac{y^2}{\sigma_y^2}\right)\right] \exp(j 2\pi f x) \\
&= g^r(x, y) + j g^i(x, y),
\end{aligned}
\tag{1.2}
$$

where $g^r(x, y)$ is the real Gabor filter as in Equation 1.1, and $g^i(x, y)$ is the imaginary component of the complex Gabor filter, given by

$$g^i(x, y) = \frac{1}{2\pi\sigma_x\sigma_y} \exp\left[-\frac{1}{2}\left(\frac{x^2}{\sigma_x^2} + \frac{y^2}{\sigma_y^2}\right)\right] \sin(2\pi f x) .$$

Analogous to the real Gabor filter, complex Gabor kernels at other angles can be obtained by rotating $g^c(x, y)$ over the range $[-\pi/2, \pi/2]$. The same design rules as described on page 7 apply for the selection of the parameters σ_x, σ_y, and f. Figure 1.4 illustrates the imaginary part of the complex Gabor filter, for $\tau = 4$ and $l = 8$.

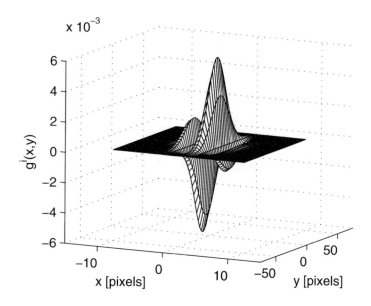

Figure 1.4: Imaginary part of the complex Gabor filter, with $\tau = 4$ and $l = 8$. Reproduced with permission from F. J. Ayres and R. M. Rangayyan. "Design and performance analysis of oriented feature detectors". Journal of Electronic Imaging, 16(2):12 pages, April 2007. article number 023007. © SPIE

The imaginary part of the complex Gabor filter approximates the Hilbert transform of the real Gabor filter; hence, the complex Gabor filter acts as a quadrature pair. Let $W_k^r(x, y)$ be the result of filtering a given image $I(x, y)$ with the real Gabor filter $g^r(x, y)$, let $W_k^i(x, y)$ be the result of filtering $I(x, y)$ with the imaginary component $g^i(x, y)$ of the complex Gabor filter, and let $W_k^c(x, y)$ be the result of filtering $I(x, y)$ with the complex Gabor filter $g^c(x, y)$. From Equation 1.2, it is evident that the following relationship holds: $|W_k^c(x, y)|^2 = |W_k^r(x, y)|^2 + |W_k^i(x, y)|^2$. The orientation field, computed with the complex Gabor filter bank, is given by the angle field $\phi^c(x, y)$, with

$$\phi^c(x, y) = \alpha_{k_{max}} \quad \text{where} \quad k_{max} = \arg\{\max_k[|W_k^c(x, y)|]\} ,$$

and by the magnitude of the output of the complex Gabor filter at the optimal orientation $E^c(x, y) = |W^c_{k_{max}}(x, y)|$. The filtering operation is implemented in the frequency domain, in the present analysis. Analogous to the quadrature steerable filter, the complex Gabor filter detects oriented features of both positive and negative contrast.

1.2.3   LINE OPERATOR

In a recent article, Zwiggelaar et al. [16] compared several methods for the detection of linear structures in mammographic images, namely: steerable filter [26], orientated bins [37], ridge detector [38], and line operator [22]. The line operator was shown to have the best detection performance among the oriented feature detectors investigated. The oriented feature detection methods investigated by Zwiggelaar et al. operate in a multiscale mode: the original image is decomposed in a Gaussian pyramid [39], and the methods are applied to each level of the pyramid.

The basic line operator kernel is sensitive to horizontal lines: detection of lines at an arbitrary orientation is achieved by rotating the basic line operator kernel. Let $H(x, y)$ be the average of N_L pixels along a horizontal line segment centered at (x, y), and $B(x, y)$ be the average of the pixel values inside a square box of width N_L, centered at (x, y) and aligned with the Cartesian axes. The line operator kernel is then given by $L(x, y) = H(x, y) - B(x, y)$. In this analysis, $N_L = 5$ pixels is employed. Figure 1.5 illustrates the basic line operator kernel.

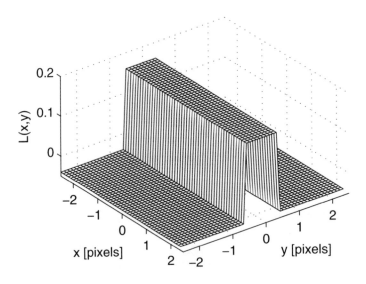

Figure 1.5: Basic line operator kernel, aligned with the x axis. Reproduced with permission from F. J. Ayres and R. M. Rangayyan. "Design and performance analysis of oriented feature detectors". Journal of Electronic Imaging, 16(2):12 pages, April 2007. article number 023007. © SPIE

Let $I(x, y)$ be the image containing the lines to be detected, and $L_k(x, y)$ be the line operator kernel rotated to the angle $\alpha_k = -\pi/2 + \pi k/K, k = 0, 1, \cdots, (K-1)$, with $K = 18$ in the present analysis. Let $W_k^l(x, y)$ be the result of filtering $I(x, y)$ with $L_k(x, y)$. The orientation of the line detected at pixel (x, y) is denoted by $\phi^l(x, y)$, and is obtained as

$$\phi^l(x, y) = \alpha_{k_{\max}} \quad \text{where} \quad k_{\max} = \arg\{\max_k [W_k^l(x, y)]\} \,,$$

and the amplitude of the result of the line operator filter is given by $E^l(x, y) = W_{k_{\max}}^l(x, y)$. The filter is sensitive to the presence of oriented features of positive contrast only.

The filtering operation is implemented as a circular convolution, in order to match the filtering strategy employed for the steerable filters and Gabor filters [34].

Contrary to the steerable filters and the Gabor filters, the line operator does not provide a parameter for scaling. Multiscale analysis using the line operator is performed by decomposing $I(x, y)$ into a Gaussian pyramid [39] and applying the line operator at each level of the pyramid. Two operations are defined in the framework of Gaussian pyramids: reduction (low-pass filtering followed by decimation by a factor of two) and expansion (the image is augmented with the interleaving of zeros and subsequent filtering, expanding the resolution by a factor of two). The filtering steps are implemented with circular convolution, analogous to the line-operator filter. The Gaussian pyramid of a given image $I(x, y)$ is a collection of images $I_p(x, y)$ obtained by successive reduction, where p denotes the number of reductions applied. The expanded images $I_p^e(x, y)$ are obtained through the expansion of $I_p(x, y)$ to the original resolution of $I(x, y)$.

The scale of the line operator is defined as $\lambda = 2^p$, where p indicates that the line operator is applied to the reduced image $I_p(x, y)$. The magnitude of the result of the line operator applied to $I_p(x, y)$ is expanded to the original resolution, in order to obtain the final magnitude $E^l(x, y)$. The line orientation produced by the application of the line operator to $I_p(x, y)$ is expanded to the original resolution using nearest-neighbor interpolation. (Linear interpolation is not applicable to angular quantities, as it would lead to inconsistent results. For instance, the average of 0° and 180° is 90°, indicating a vertical orientation that is incompatible with the horizontal orientations being averaged.)

1.2.4 FREQUENCY-DOMAIN ANALYSIS

All of the oriented feature detectors discussed in this chapter can be scaled and rotated. In addition, the real and the complex Gabor filters permit the modification of the aspect ratio of the filter, through the elongation factor l.

Figure 1.6 exemplifies the effect of the different design parameters in the real Gabor filter, thus illustrating the scaling and rotation phenomena common to all filters, and the effect of the elongation factor on the (real and complex) Gabor filters. Figure 1.6 is described as follows:

- Figures 1.6a and 1.6e show the impulse response of a Gabor filter and its Fourier transform magnitude (frequency response), respectively.

- In Figure 1.6b, the Gabor filter of Figure 1.6a is stretched in the x direction, by increasing the elongation factor l. Observe that the Fourier spectrum of the new Gabor filter, shown in Figure 1.6f, is compressed in the horizontal direction.

- The Gabor filter shown in Figure 1.6c is obtained by increasing the parameter τ of the original Gabor filter, thus enlarging the filter in both the x and y directions. Correspondingly, the Fourier spectrum of the enlarged filter, shown in Figure 1.6g, is shrunk in both the vertical and horizontal directions.

- The effect of rotating the Gabor filter by 30° counterclockwise is displayed in Figures 1.6d and 1.6h, that show the rotated impulse response and its corresponding Fourier spectrum, respectively.

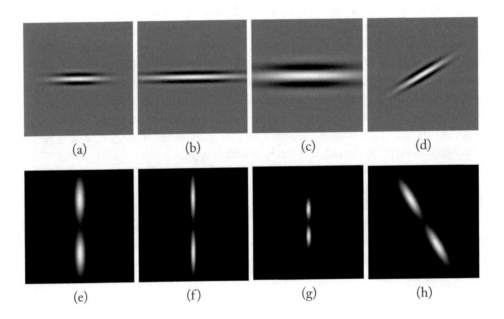

Figure 1.6: Effect of the different parameters of the Gabor filter. (a) Example of the impulse response of a Gabor filter. (b) The parameter l is increased: the Gabor filter is elongated in the x direction. (c) The parameter τ is increased: the Gabor filter is enlarged in the x and y directions. (d) The angle of the Gabor filter is modified. Figures e–f correspond to the magnitude of the Fourier transforms of the Gabor filters in a–d, respectively. The $(0, 0)$ frequency component is at the center of the spectra displayed.

A comparison between the impulse response and the magnitude of the frequency response for each oriented feature detector described in the present chapter is presented in Figure 1.7. It can be observed from an analysis of the frequency response of each filter that the Gabor filters (real and complex) are narrowband filters, whereas the remaining filters have coarser angular selectivity.

Oriented feature detector	Impulse response	Frequency response (magnitude)

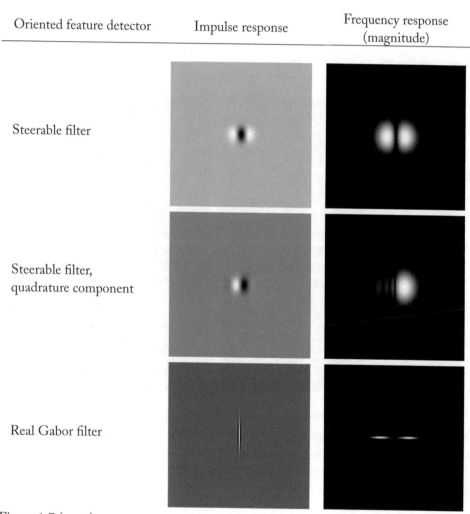

Steerable filter

Steerable filter,
quadrature component

Real Gabor filter

Figure 1.7 (cont.)

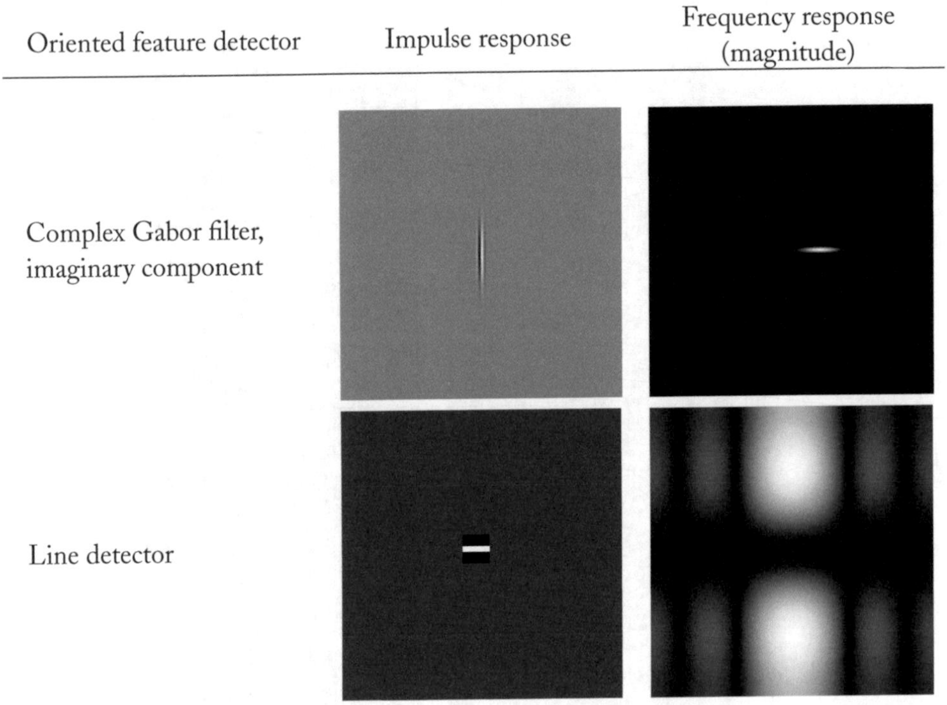

Oriented feature detector	Impulse response	Frequency response (magnitude)
Complex Gabor filter, imaginary component		
Line detector		

Figure 1.7: Impulse response and magnitude component of the frequency response of each oriented feature detectors. The frequency response shown with the quadrature component of the quadrature-pair steerable filter corresponds to the frequency response of the quadrature pair. The frequency response shown with the impulse response of the imaginary component of the complex Gabor filter corresponds to the frequency response of the complex Gabor filter.

1.3 PERFORMANCE ANALYSIS OF THE ORIENTED FEATURE DETECTORS

This section presents a comparison of the detection performance and orientation accuracy of the five oriented feature detectors presented in Section 1.2. The detection performance is defined in terms of the ability of each filter to detect linear structures in the presence of noise and imprecision in the specification of scale. The orientation accuracy is given in terms of the cumulative angle error for the pixels belonging to the lines in the test pattern used.

1.3.1 PERFORMANCE IN TERMS OF DETECTION AND ANGULAR ACCURACY

A test image of size 512×512 pixels was employed in the investigation of the detection performance and orientation accuracy of each oriented feature detector. The test pattern includes 34 line segments oriented at equally spaced angles, distributed radially around a circle with a radius of 115 pixels. Each line segment has a length of 115 pixels and a width of two pixels. The background of the test image was set to 0.4, and the line segments were set to 0.6, in the normalized scale [0, 1] for gray levels. The test pattern was corrupted with various levels of Gaussian noise (standard deviation range: $\sigma_\eta \in [0, 0.5]$, in steps of 0.05), allowing for the analysis of the robustness of each filter to noise. Figure 1.8a shows the test pattern corrupted by Gaussian noise of standard deviation $\sigma_\eta = 0.1$.

The number of line segments was chosen as 34 in order to prevent synchronization between the angles of the line segments and the angles of the kernels of the Gabor filter banks (real and complex) or the line-operator filter banks. Over the half-closed angular interval [0, 180) degrees, we have 17 line segments and 18 kernels in each filter bank. As a consequence, the absolute angular error between the angle of a given line segment and the closest kernel angle in the filter bank will evenly span the range [0, 5) degrees. The same reasoning applies to the line segments oriented at angles in the half-closed angular interval [180, 360) because the arithmetic of orientation angles follows a congruence with modulo 180 degrees.

The orientation field of the test image was obtained, for each level of noise, and for different values of the corresponding scale parameter with each filter. Figure 1.8 shows the orientation field magnitude obtained with each of the five oriented feature detectors for the scale parameters specified in the caption.

For each filter, a pixel was considered to be part of a linear structure if the magnitude of the orientation field at that pixel exceeded a given threshold. In this manner, receiver operating characteristics (ROC) analysis [40] can be performed to investigate the detection performance of each filter (see Appendix B). The detection performance measure was defined as the area under the ROC curve [40], denoted by AUC. The value of AUC was obtained for each noise level, and for several values of the scale parameter of each filter; the results are shown as topographic maps in Figure 1.9.

The definition of the scale parameter varies across the oriented feature detectors presented in this chapter, introducing a degree of freedom in multiscale comparison of the different detectors:

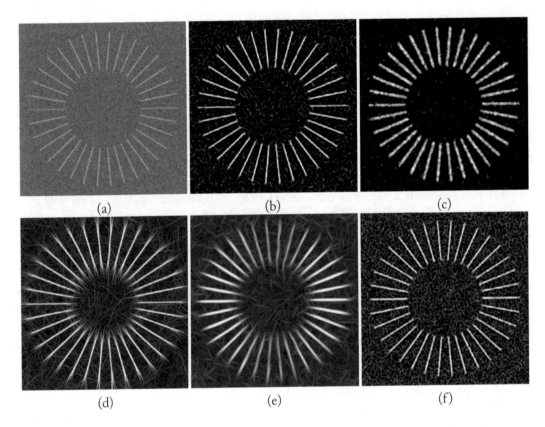

Figure 1.8: Orientation field magnitude obtained with each oriented feature detector. (a) Test pattern, corrupted with Gaussian noise ($\sigma_\eta = 0.1$). (b) Steerable filter ($\sigma_s = 2.9$ pixels). (c) Quadrature steerable filter ($\sigma_s = 2.1$ pixels). (d) Real Gabor filter ($\tau = 8.5$ pixels). (e) Complex Gabor filter ($\tau = 5.5$ pixels). (f) Line operator ($\lambda = 2$ pixels). Reproduced with permission from F. J. Ayres and R. M. Rangayyan. "Design and performance analysis of oriented feature detectors". Journal of Electronic Imaging, 16(2):12 pages, April 2007. article number 023007. © SPIE

the scale measurement of each detector may differ from the remaining detectors by a multiplicative factor. Therefore, the base-ten logarithm of the scale parameter is employed in the present analysis, instead of the linear scale, in order to allow straightforward comparison of the different detectors. The logarithm operation converts any multiplicative factor into a summation constant, and comparisons of interval length in the scale parameter are not influenced by this constant. In particular, it is possible to integrate the detection performance of each detector (given by AUC, as previously mentioned) with respect to the logarithm of the scale parameter.

The following characteristics were observed in the results of all of the oriented feature detectors investigated in this chapter:

- The value of AUC decreases with the noise level. A slower decay in AUC versus noise denotes greater robustness to noise.

- For a given level of noise, there is an optimal range of values for the scale parameter associated with a high AUC value. This range can be interpreted as an indicator of robustness to imprecision in the specification of scale in the design of the filter, in relation to the actual width of the features in the given image. Conversely, for a single-scale detector, the range of optimal values is associated with the variation of characteristic widths of oriented features for which the oriented feature detector can provide a high detection capability.

A measure to represent the combined robustness to noise and scale of each detector is proposed, as the area of the domain (noise level, $\log_{10}$ of the scale parameter) for which $AUC > 0.9$; see Figure 1.9. The filter that has the largest area under the topographic line corresponding to $AUC = 0.9$ is the most robust among the filters evaluated. Table 1.1 presents the robustness measure of each detector.

In order to illustrate better the robustness of each oriented feature detector to imprecision in the specification of scale, the standard deviation (σ_η) of the noise in the test image was set to 0.1, and the area under the ROC curve AUC was analyzed as a function of scale only, for each detector. The following parameters were obtained for each oriented feature detector, and the results are shown in Table 1.1:

- the best scale;

- the range of scale (in pixels) in which $AUC > 0.9$, called the detection range; and

- the equivalent detection range, in millimeters, if the center of the scale range as defined above is normalized to the width of 1 mm.

In order to evaluate the angular precision of the detectors, the cumulative angular error was obtained at the best scale (shown in Table 1.1), and with a noise standard deviation of $\sigma_\eta = 0.1$ for the test image, as shown in Figure 1.8a. Figure 1.10 displays the results of the cumulative angular error analysis, i.e., the fraction of the positive (line) pixels whose angular error is less than a given angular tolerance. It is observed that the real and complex Gabor filters present the best accuracy:

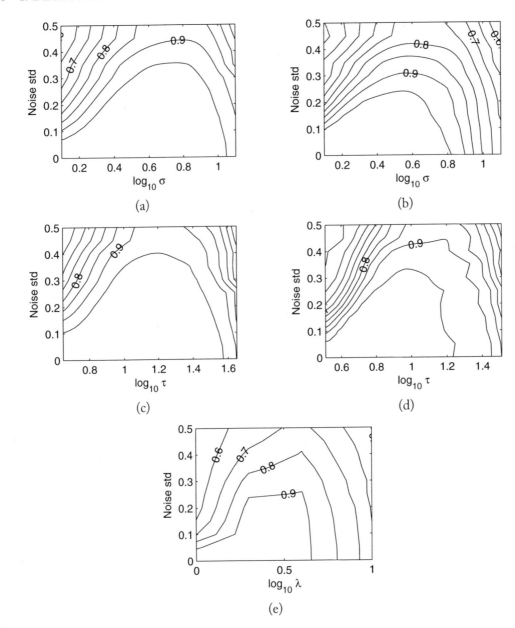

Figure 1.9: Topographic maps showing AUC for the oriented feature detectors studied, at various levels of noise and scale (std = standard deviation). (a) Steerable filter. (b) Quadrature steerable filter. (c) Real Gabor filter. (d) Complex Gabor filter. (e) Line operator. Reproduced with permission from F. J. Ayres and R. M. Rangayyan. "Design and performance analysis of oriented feature detectors". Journal of Electronic Imaging, 16(2):12 pages, April 2007. article number 023007. © SPIE

Table 1.1: Robustness measure, best scale, and detection range for each oriented feature detector. The best scale and the detection range values were obtained at a noise standard deviation of $\sigma_\eta = 0.1$ in the test image. Noise in the range of $[0, 0.5]$ in standard deviation was used to derive the measure of robustness. Reproduced with permission from F. J. Ayres and R. M. Rangayyan. "Design and performance analysis of oriented feature detectors". Journal of Electronic Imaging, 16(2):12 pages, April 2007. article number 023007. © SPIE

Detector	Robustness	Best scale [pixels]	Detection range [pixels]	Detection range [mm]
Steerable	0.36	2.9	1.2 − 14.2	0.2 − 1.8
Quadrature steerable	0.22	2.1	1.1 − 7.4	0.3 − 1.7
Real Gabor	0.46	8.5	3.3 − 42.2	0.1 − 1.9
Complex Gabor	0.34	5.5	3.4 − 26.6	0.2 − 1.8
Line operator	0.37	2.0	1.5 − 11.2	0.2 − 1.8

almost 100% of the line pixels (pixels belonging to the true line pattern in the test image) in the orientation field generated by these filters present an angular error less than 10 degrees.

1.3.2   ANALYSIS OF PERFORMANCE WITH MULTISCALE FEATURES

The detection range of each filter, in terms of capturing features of varying width or scale, was analyzed with the following experiment. A test image of 4096×512 pixels was created, consisting of 32 bright lines (of intensity 0.6) on a relatively dark background (of intensity 0.4), as shown in Figure 1.11a. The lines are vertically oriented, with the length (or height) of 212 pixels, and widths varying from one to 32 pixels in steps of one pixel. This test image will be referred to as the multiscale test image in the following analysis. Gaussian noise of standard deviation $\sigma_\eta = 0.1$ was added to the multiscale test image (see Figure 1.11b). The scale parameter of each oriented feature detector was set as follows:

- The geometric mean between the minimum line width (one pixel) and the maximum line width (32 pixels) in the multiscale test image is 5.66. This value represents the center of the line width range in the multiscale test image.

- The detection range results presented in Table 1.1 were obtained using a line width of two pixels, as described in Section 1.3.1. The ratio between the width of the center line in the multiscale test image and the line width in the first test image in Figure 1.8a is 2.83.

- The scale parameter of each oriented feature detector is set as the geometric mean of the detection range of each oriented feature detector, as given in Table 1.1, times the rescaling factor of 2.83, computed above.

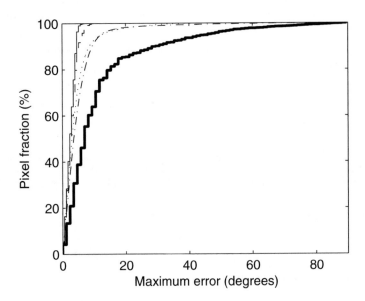

Figure 1.10: Cumulative angular error for each oriented feature detector: steerable filter (dotted line); quadrature steerable filter (dash-dot line); real Gabor filter (dashed line); complex Gabor filter (solid line); line operator (bold solid line). The curves associated with the real and complex Gabor filters overlap significantly. Reproduced with permission from F. J. Ayres and R. M. Rangayyan. "Design and performance analysis of oriented feature detectors". Journal of Electronic Imaging, 16(2):12 pages, April 2007. article number 023007. © SPIE

Figures 1.11c to 1.11g show the resulting magnitude images from the application of each oriented feature detector. In all cases, the area under the ROC curve was greater than 0.9. Observe that all of the oriented feature detectors provide a stronger detection at the middle of the multiscale test image than at its ends, as expected. For thinner lines, the oriented feature detectors exhibit a less significant response than those for the lines in the middle of the test image. The application of the oriented feature detectors to thicker lines also generated less significant responses along the core of the line features, with the exception of the steerable filter (Figure 1.11c). It can be observed that, in the case of the quadrature steerable filter and the complex Gabor filter, only the edges of the thicker line features have been identified as oriented features.

1.3.3   COMPUTATIONAL TIME

The execution time was measured for each oriented feature detector and averaged over 10 repetitions, on a Dell Precision 360 workstation (3.20 GHz Pentium 4 processor, 2 MB of cache, and 2 GB of RAM); the results are shown in Table 1.2. The line operator has the lowest execution time, followed

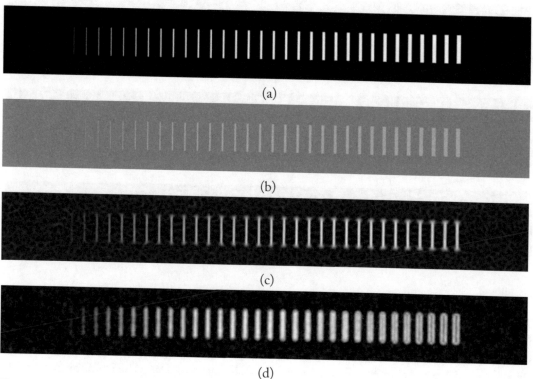

(a)

(b)

(c)

(d)

Figure 1.11 (cont.)

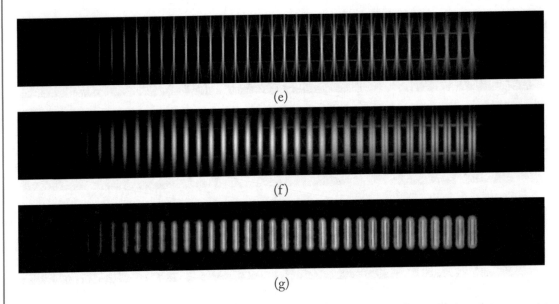

(e)

(f)

(g)

Figure 1.11: Application of the oriented feature detectors to a test pattern of lines of varying width. (a) Test pattern. (b) Test pattern with Gaussian noise added ($\sigma_\eta = 0.1$). (c) Steerable filter ($\sigma_s = 11.67$ pixels). (d) Quadrature steerable filter ($\sigma_s = 8.06$ pixels). (e) Real Gabor filter ($\tau = 33.38$ pixels). (f) Complex Gabor filter ($\tau = 26.90$ pixels). (g) Line operator ($\lambda = 8$ pixels). All images are of size 4096×512 pixels. Reproduced with permission from F. J. Ayres and R. M. Rangayyan. "Design and performance analysis of oriented feature detectors". Journal of Electronic Imaging, 16(2):12 pages, April 2007. article number 023007. © SPIE

by the steerable filter, the quadrature steerable filter, the real Gabor filter, and the complex Gabor filter.

Table 1.2: Mean execution time for each oriented feature detector, when applied to the 512×512 test pattern shown in Figure 1.8a. The execution time was averaged over 10 repetitions on a Dell Precision 360 workstation (3.20 GHz Pentium 4 processor, 2 MB of cache, and 2 GB of RAM).

Detector	Execution time [s]
Steerable	1.36
Quadrature steerable	3.65
Real Gabor	5.00
Complex Gabor	6.23
Line operator	0.75

1.4 DISCUSSION

The aim of the analysis described in this chapter was to develop design criteria and conduct performance analysis of five oriented feature detectors, in terms of their ability in the detection of oriented features and accuracy in the determination of the angle of the oriented features of interest. Performance measures were derived and analyzed in the presence of noise and imprecision in the specification of scale.

All of the oriented feature detectors investigated in this chapter are based on filter banks consisting of filtering kernels that are sensitive to the presence of oriented features. Each filter presents a positive lobe that is elongated at a certain orientation and negative sidelobes in order to highlight the presence of linear features at a particular orientation. The results of filtering for each filter bank are combined to form two maps: the magnitude map, which indicates the intensity of the oriented feature at each pixel, and the angle map, which indicates the orientation of the feature at each pixel. The magnitude and angle maps are together referred to as the orientation field of the image under analysis.

1.4.1 ROBUSTNESS TO NOISE AND SCALE

The detection performance of the oriented feature detectors in the presence of noise and imprecision in scale is shown in Figure 1.9. It is observed that all of the oriented feature detectors studied present a reduction in AUC with increasing noise, as expected. The real Gabor filter is the most robust to noise, presenting a large range of noise over which the detection performance is greater than 0.9. This result can be explained in terms of the spatial extent of the filters investigated: the main lobe of the real Gabor filter has a larger length-to-width ratio than that of the other filters, thus enhancing more effectively the signal-to-noise ratio after filtering.

The robustness to imprecision in scale specification is also shown in Figure 1.9; it is noticed that, for a given level of noise in the test image, each oriented feature detector has an optimal range of scales in which the detector presents good detection performance (defined as $AUC > 0.9$). Table 1.1 shows the optimal scale range of each oriented feature detector, for a noise standard deviation level of 0.1 in the test image; conversion of the optimal scale range to millimeters is also given for easier comparison of the oriented feature detectors. The optimal scale range was converted to millimeters in such way that the oriented feature width of 1 mm is placed at the center of the optimal scale ranges. It is observed that all filters present similar optimal scale ranges (in millimeters); the real Gabor filter presents the largest range.

It must be noted that the optimal scale range of all oriented feature detectors can be enhanced with the adoption of a multiscale strategy: each oriented feature detector can be applied to the image under analysis at different values of the scale parameter, resulting in an orientation field for each scale. The final orientation field is then obtained as follows: the magnitude of the final orientation field, at a given pixel, is the largest magnitude across the set of orientation fields that are obtained for each scale; the final angle is taken as the angle of the orientation field at the scale that provided the largest magnitude response. Nevertheless, the analysis of the optimal scale range is relevant: given a range of characteristic widths of several oriented features to be detected in an image, the analysis of the optimal scale range as above permits the appropriate selection of the scale parameter values, in the multiscale operation of each oriented feature detector. The set of scale parameter values must be chosen in such a manner that the union of the optimal scale ranges, associated with each scale chosen, covers the given range of the characteristic widths of the oriented features of interest.

The effect of noise and scale imprecision in the detection capability of each oriented feature detector may be summarized in a single combined robustness measure, as shown in Table 1.1 and described in Section 1.3. It can be observed that the real Gabor filter has the best combined robustness measure, followed by the line operator, the steerable filter, the complex Gabor, and quadrature steerable filters. As mentioned in the preceding discussion, all filters have comparable optimal scale ranges, and the real Gabor filter has higher robustness to noise. Therefore, the ranking of the oriented feature detectors according to the combined robustness measure is consistent with the preceding discussion.

The superior performance of the real Gabor, the line operator, and the steerable filters is due to the fact that the test pattern used to study the oriented feature detectors consists of line segments of positive contrast; therefore, the filters have an intrinsic advantage in the detection of the lines in the test pattern. The real Gabor, the line operator, and the steerable filters are recommended when all oriented features to be detected in an image are of positive contrast. (Detection of oriented features of negative contrast in an image can be achieved by detecting oriented features of positive contrast in the negative of the given image.) In many real-life applications, the contrast of the oriented features to be detected is known in advance; for instance, the presence of malignant tumors or architectural distortion in mammograms may result in spiculations, which are oriented features of positive contrast [41]. When the contrast of the oriented features to be detected is not known, or if the given image may exhibit oriented features of both positive and negative contrast, the real Gabor, the line operator, and steerable filters may still be used: the oriented features must be detected in both the given image and in its negative, and the corresponding orientation fields must be combined. An alternative to the detection of oriented features of unknown contrast is the application of the complex Gabor and the quadrature steerable filters, albeit at a higher computational cost.

1.4.2   ANGULAR ACCURACY

The angular precision of the oriented feature detectors is illustrated in Figure 1.10. It can be observed that the Gabor filters have the best accuracy among the oriented feature detectors studied in this analysis. The line operator has the poorest orientation accuracy. These results are a consequence of the length-to-width aspect ratio of the main lobes of the filters. The Gabor filters used in the present analysis have the highest aspect ratio; hence, the oriented Gabor kernel corresponding to the optimal orientation prevails over the remaining kernels in the Gabor filter bank in terms of the magnitude of the output, resulting in less ambiguity when deciding upon the orientation of the feature at a given pixel. The line operator of Dixon and Taylor [22] and steerable filters do not provide parameters to modify the length-to-width aspect ratio.

1.4.3   COMPUTATIONAL SPEED

The line operator has the highest computational speed, as seen in Table 1.2, due to the efficient implementation of the filter as a convolution operator. The Gabor filters require the highest computational effort: both the real and complex Gabor filter banks used in the present analysis (with the synthetic test images) include 18 filters, which were implemented as frequency-domain filters instead of convolution operations in the space domain.

1.5   APPLICATION OF GABOR FILTERS TO THE DETECTION OF CURVILINEAR STRUCTURES IN MAMMOGRAMS

Oriented features present in mammographic images are related to various normal structures in the breast (such as vessels, ducts, and fibroglandular tissue), and to abnormal elements that may be associated with the presence of breast cancer (such as spicules in a spiculated tumor or architectural distortion; see Appendix A). An example of the orientation field of a mammographic image is given in Figures 1.12, 1.13, and 1.14. Figure 1.12 shows a mammographic image from the Mini-MIAS database [42] (image "mdb168"). The orientation field of the mammographic image was obtained using a bank of real Gabor filters, as described in Section 1.2.2, with $K = 180$. Figures 1.13 and 1.14

show the magnitude and angle components of the orientation field of the mammographic image. It can be observed that the oriented features present in the mammographic image are emphasized in the magnitude component of the orientation field (Figure 1.13), and the orientation angle depicted in Figure 1.14 agrees with the appearance of the oriented features present in the mammographic image.

The organization of oriented features in mammograms may be used to detect abnormal conditions, as indicated in the two examples described below:

- Architectural distortion is a subtle mammographic abnormality that is characterized by distortion of the normal structure of the breast, with no mass visible, and may appear as a set of spicules radiating from a central point, or as a focal retraction at the edge of the fibroglandular disk. Methods for the detection of architectural distortion are presented in Chapter 4.

- Bilateral asymmetry occurs when the left and right breasts differ in their mammographic appearance. This mammographic finding has been linked to an increased probability of development of breast cancer [43]. Ferrari *et al.* [30] used Gabor filters and multiscale analysis to obtain the orientation field of mammographic images, from which histograms of orientation angles (or rose diagrams) were obtained. A set of features related to the statistical distribution of orientation angles was obtained, and used to detect the occurrence of bilateral asymmetry. The authors reported a classification accuracy of 74.4%.

Appendix A presents a further discussion on computer-aided analysis of mammographic images.

1.6   REMARKS

Based upon the results presented in this chapter, it is evident that the real Gabor filter presents the best combination of detection performance and angular accuracy. The line operator has the highest computational speed, followed by the steerable and the quadrature steerable filters; the Gabor filters have the highest computational requirement among the oriented feature detectors analyzed in this chapter. Despite the high computational speed of the line operator, the steerable filter is recommended where computational speed is important, due to its significantly improved angular accuracy and comparable detection performance. The real Gabor filter is recommended when high detection performance and angular accuracy are required, as in the case of noisy images or when the presence of oriented features is subtle. We have found the real Gabor filter to perform well in the detection of architectural distortion in mammograms [14]; details of this application are presented in Chapter 4.

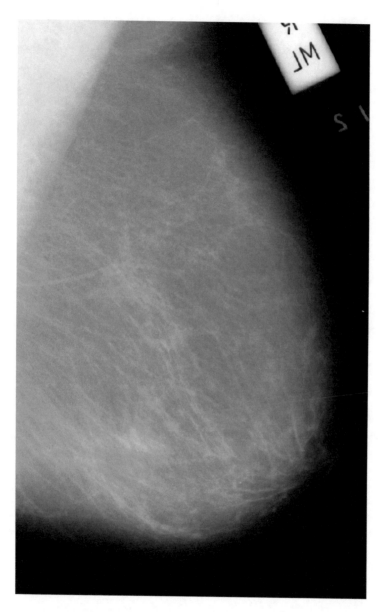

Figure 1.12: A normal mammographic image (mdb168) from the Mini-MIAS database. Image size: 646 × 1024 pixels. Pixel resolution is 200 μm/pixel. Reproduced with permission from F. J. Ayres and R. M. Rangayyan. "Design and performance analysis of oriented feature detectors". Journal of Electronic Imaging, 16(2):12 pages, April 2007. article number 023007. © SPIE

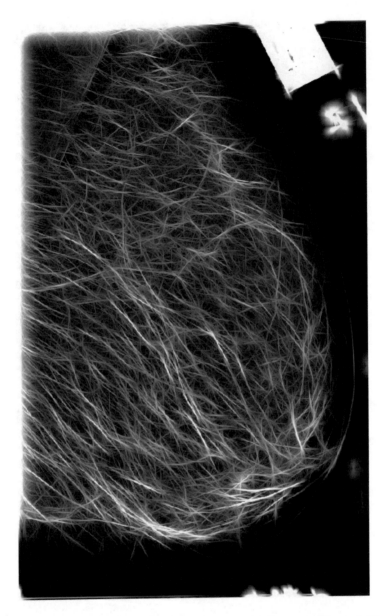

Figure 1.13: Orientation field magnitude for the mammographic image in Figure 1.12, obtained using a bank of 180 real Gabor filters. Reproduced with permission from F. J. Ayres and R. M. Rangayyan. "Design and performance analysis of oriented feature detectors". Journal of Electronic Imaging, 16(2):12 pages, April 2007. article number 023007. © SPIE

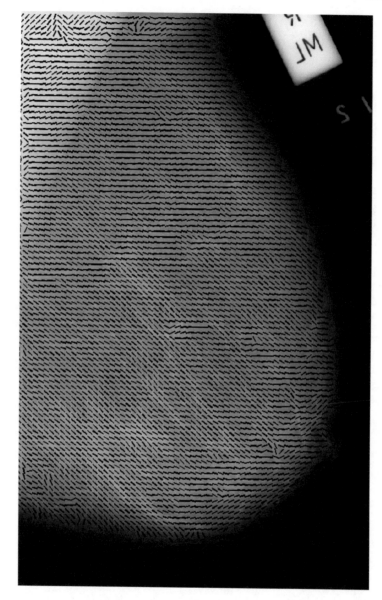

Figure 1.14: Orientation field angle superimposed on the mammographic image in Figure 1.12. The orientation field was obtained using a bank of 180 Gabor filters. Line segments are drawn every 10 pixels, indicating the orientation field angle. The orientation field angle has been smoothed using the procedure described in Section 4.1.1. Reproduced with permission from F. J. Ayres and R. M. Rangayyan. "Design and performance analysis of oriented feature detectors". Journal of Electronic Imaging, 16(2):12 pages, April 2007. article number 023007. © SPIE

CHAPTER 2

Analysis of Oriented Patterns Using Phase Portraits

Consider a system of two linear, first-order, differential equations. A diagram that depicts the possible trajectories of the state variables, for different initialization values, is called the phase portrait diagram of the system. The appearance of the phase portrait diagram can be categorized as node, saddle, or spiral.

Rao and Jain [44] (see also Rao [45]) developed a method for the analysis of images displaying oriented texture, which relies on the association of the oriented texture pattern present in an image with the appearance of a phase portrait diagram (and the corresponding parameters of the system of differential equations). In this chapter, the method of phase portraits is described and illustrative examples are given. We employ the phase portrait method for the detection of architectural distortion in mammograms.

2.1 PHASE PORTRAITS

Phase portraits provide an analytical tool to study systems of first-order differential equations [46]. Let $p(t)$ and $q(t)$, $t \in \mathbb{R}$, denote two differentiable functions of time t, related by a system of first-order differential equations as

$$\begin{aligned} \dot{p}(t) &= F[p(t), q(t)] \\ \dot{q}(t) &= G[p(t), q(t)], \end{aligned} \qquad (2.1)$$

where the dot above the variable indicates the first-order derivative of the function with respect to time, and F and G represent functions of p and q. Given initial conditions $p(0)$ and $q(0)$, the solution $(p(t), q(t))$ to Equation 2.1 can be viewed as a parametric trajectory of a hypothetical particle in the pq plane, placed at $(p(0), q(0))$ at time $t = 0$, and moving through the pq plane with velocity $(\dot{p}(t), \dot{q}(t))$. The pq plane is referred to as the *phase plane* of the system of first-order differential equations. The path traced by the hypothetical particle is called a *streamline* of the vector field $(\dot{p}, \dot{q})$. The *phase portrait* is a graph of the possible streamlines in the phase plane. A *fixed point* of Equation 2.1 is a point in the phase plane where $\dot{p}(t) = 0$ and $\dot{q}(t) = 0$: a particle left at a fixed point remains stationary.

When the system of first-order differential equations is linear, Equation 2.1 assumes the form

$$\begin{pmatrix} \dot{p}(t) \\ \dot{q}(t) \end{pmatrix} = \mathbf{A} \begin{pmatrix} p(t) \\ q(t) \end{pmatrix} + \mathbf{b}, \qquad (2.2)$$

where $\mathbf{A}$ is a 2×2 matrix and $\mathbf{b}$ is a 2×1 column matrix (a vector). In this case, there are only three types of phase portraits: node, saddle, and spiral [46]. The type of phase portrait can be determined from the nature of the eigenvalues of $\mathbf{A}$, as shown in Table 2.1. The eigenvalues of $\mathbf{A}$ are given by

$$\lambda_1 = \frac{tr(\mathbf{A})}{2} + \frac{\sqrt{[tr(\mathbf{A})]^2 - 4\,det(\mathbf{A})}}{2} \qquad (2.3)$$

and

$$\lambda_2 = \frac{tr(\mathbf{A})}{2} - \frac{\sqrt{[tr(\mathbf{A})]^2 - 4\,det(\mathbf{A})}}{2}, \qquad (2.4)$$

where $tr(\mathbf{A})$ is the trace of matrix $\mathbf{A}$ and $det(\mathbf{A})$ is the determinant of $\mathbf{A}$.

The center (p_0, q_0) of the phase portrait is given by the fixed point of Equation 2.2:

$$\begin{pmatrix} \dot{p}(t) \\ \dot{q}(t) \end{pmatrix} = 0 \Rightarrow \begin{pmatrix} p_0 \\ q_0 \end{pmatrix} = -\mathbf{A}^{-1}\mathbf{b}. \qquad (2.5)$$

Solving Equation 2.2 yields a linear combination of complex exponentials for $p(t)$ and $q(t)$, whose exponents are given by the eigenvalues of $\mathbf{A}$ multiplied by the time variable t. Table 2.1 illustrates the streamlines obtained by solving Equation 2.2 for a node, a saddle, and a spiral phase portrait: the solid lines indicate the movement of the $p(t)$ and the $q(t)$ components of the solution, and the dashed lines indicate the streamlines. The formation of each phase portrait type is explained as follows:

- *Node* — The components $p(t)$ and $q(t)$ are exponentials that either simultaneously converge to, or diverge from, the fixed-point coordinates p_0 and q_0. This condition occurs when the eigenvalues of $\mathbf{A}$ have the same sign. From Equations 2.3 and 2.4, we conclude that a node pattern is obtained when

$$0 < det(\mathbf{A}) < \frac{[tr(\mathbf{A})]^2}{4}.$$

- *Saddle* — The components $p(t)$ and $q(t)$ are exponentials; while one of the components (either $p(t)$ or $q(t)$) converges to the fixed point, the other diverges from it. A saddle pattern occurs when the eigenvalues of $\mathbf{A}$ are real and have opposite signs. This condition is achieved if

$$det(\mathbf{A}) < 0.$$

- *Spiral* — The components $p(t)$ and $q(t)$ are exponentially modulated sinusoidal functions; the resulting streamline forms a spiral curve. The eigenvalues of $\mathbf{A}$ form a pair of complex conjugate numbers with nonzero imaginary parts. In this case, we have

Table 2.1: Phase portraits for a system of linear first-order differential equations. Solid lines indicate the movement of the $p(t)$ and the $q(t)$ components of the solution; dashed lines indicate the streamlines. Reproduced with permission from F. J. Ayres and R. M. Rangayyan. "Characterization of architectural distortion in mammograms". IEEE Engineering in Medicine and Biology Magazine, 24(1):59–67, January 2005. © IEEE

Phase portrait type	Eigenvalues	Streamlines	Appearance of the orientation field
Node	Real eigenvalues of same sign		
Saddle	Real eigenvalues of opposite sign		
Spiral	Complex eigenvalues		

$$det(\mathbf{A}) > \frac{[tr(\mathbf{A})]^2}{4}.$$

The relationship between the type of phase portrait and the values of $det(\mathbf{A})$ and $tr(\mathbf{A})$ is illustrated in Figure 2.1.

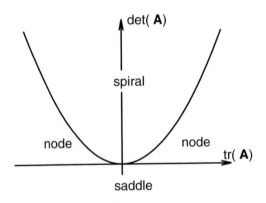

Figure 2.1: Type of phase portrait in relation to $det(\mathbf{A})$ and $tr(\mathbf{A})$.

If a model matrix $\mathbf{A}$ has determinant and trace values positioned near the interface between types of phase portrait in Figure 2.1, the assignment of a definite type of phase portrait to $\mathbf{A}$ in a real application cannot be done reliably.

2.2   ANALYSIS OF ORIENTATION FIELDS USING PHASE PORTRAITS

2.2.1   LOCAL ANALYSIS OF ORIENTATION FIELDS

The model in Equation 2.2 can be used to analyze orientation fields, as proposed by Rao and Jain [44]. Consider the following vector field model:

$$\vec{v} = \begin{pmatrix} v_x \\ v_y \end{pmatrix} = \mathbf{A} \begin{pmatrix} x \\ y \end{pmatrix} + \mathbf{b} , \tag{2.6}$$

where

$$\mathbf{A} = \begin{bmatrix} a & b \\ c & d \end{bmatrix}, \quad \mathbf{b} = \begin{bmatrix} e \\ f \end{bmatrix} . \tag{2.7}$$

The vector $\vec{v}$ is an affine function of the coordinates (x, y). A particle on the Cartesian (image) plane whose velocity is given by $\vec{v}(x, y)$ will follow a trajectory that is analogous to the time evolution of the dynamical system in Equation 2.2. Therefore, Equation 2.6 can be compared to Equation 2.2 by associating the vector $\vec{v}$ with the state velocity $(\dot{p}(t), \dot{q}(t))$, and the position (x, y) with the state $(p(t), q(t))$. The *orientation field* generated by Equation 2.6 can be defined as

$$\phi(x, y|\mathbf{A}, \mathbf{b}) = \arctan\left(\frac{v_y}{v_x}\right), \tag{2.8}$$

which is the angle of the vector $\vec{v}$ with the x axis.

Table 2.1 lists the three phase portraits and the corresponding orientation fields generated by a system of linear first-order differential equations.

The orientation field of an image $I(x, y)$ presenting oriented texture can be obtained using the real Gabor filter bank described in Section 1.2.2. Let us denote by $M(x, y)$ and $\theta(x, y)$ the magnitude and angle components of the orientation field of $I(x, y)$. At every pixel location (x, y), define the error between $\theta(x, y)$ and a synthetic orientation field $\phi(x, y|\mathbf{A}, \mathbf{b})$ as

$$r(x, y) = \sin\left(\theta(x, y) - \phi(x, y|\mathbf{A}, \mathbf{b})\right). \tag{2.9}$$

The sum of the squared errors (referred to henceforth as the error measure), weighted by the magnitude of the orientation field, is given by

$$\epsilon^2(\mathbf{A}, \mathbf{b}) = \sum_x \sum_y M(x, y) \sin^2(\theta(x, y) - \phi(x, y|\mathbf{A}, \mathbf{b})). \tag{2.10}$$

Minimizing $\epsilon^2(\mathbf{A}, \mathbf{b})$ with respect to the elements of $\mathbf{A}$ and $\mathbf{b}$ yields a set of parameters $\mathbf{A}_{opt}$ and $\mathbf{b}_{opt}$, which relate to the synthetic orientation field that best approximates the orientation field of $I(x, y)$. [Optimization methods for the minimization of $\epsilon^2(\mathbf{A}, \mathbf{b})$ are presented in Chapter 3.] The parameters $\mathbf{A}_{opt}$ and $\mathbf{b}_{opt}$ permit the determination of the type of phase portrait and the location of the fixed point of $\phi(x, y|\mathbf{A}_{opt}, \mathbf{b}_{opt})$, which provide a qualitative description of the oriented texture and its focal point, respectively, in the image under investigation.

The sine of the difference between angles, as in Equation 2.9, serves as a better measure of the difference between two orientations than the difference between the angles. For example, consider two line segments oriented at $0°$ and $179°$ with respect to a common reference. The difference between the angles is $179°$, whereas the sine of the difference is close to zero. The latter accurately reflects the fact that the line segments have nearly the same orientation.

2.2.2   ANALYSIS OF LARGE ORIENTATION FIELDS

Large orientation fields may contain complex patterns that are superpositions of several simpler patterns, resulting in the presence of multiple focal points. The approximation of real orientation fields using Equation 2.8 is only valid locally. A general strategy to extend the method for the local analysis of orientation fields (presented in Section 2.2.1) to the analysis of large orientation fields is to analyze the large orientation field at multiple locations, and to accumulate the obtained information in a form that permits the identification of the various relevant structures present in the overall orientation field.

Rao and Jain [44] proposed the following method for the analysis of large orientation fields:

1. Create three images of the same resolution as that of the image or orientation field under analysis. These three images will be referred to as *phase portrait maps*. Initialize the phase portrait maps to zero.

2. Move a sliding analysis window throughout the orientation field. For every position of the analysis window, perform the following steps:

    (a) Use the procedure described in Section 2.2.1 to find the optimal parameters $\mathbf{A}_{opt}$ and $\mathbf{b}_{opt}$ that best describe the orientation field under the analysis window.

    (b) From $\mathbf{A}_{opt}$ and $\mathbf{b}_{opt}$, determine the type of phase portrait and the fixed point location associated with the orientation field under the analysis window.

    (c) Select the phase portrait map corresponding to the phase portrait type determined above. Increment the value present at the pixel nearest to the fixed-point location. This procedure will be referred to as *vote casting*.

After all votes are cast, the phase portrait maps may be analyzed to detect the presence of patterns in the given image or orientation field. If a portion of the orientation field is comprised of orientations radiating from a central point, in a manner similar to a node pattern, it is expected that the node map will contain an agglomeration of votes close to the geometrical focus of the observed pattern. Likewise, the presence of patterns that are similar to saddle or spiral patterns will cause agglomerations of votes in the respective phase portrait maps.

2.3 ILLUSTRATIVE EXAMPLES

2.3.1 SYNTHETIC TEST IMAGES

In order to illustrate the technique described in Section 2.2.2, three synthetic test images exhibiting oriented texture were created. Each image is associated with a specific type of phase portrait. The images were created as follows:

1. A vector field $\vec{v}(x, y)$ of size 512×512 pixels was created, for a given set of parameters $\mathbf{A}$ and $\mathbf{b}$, using Equation 2.6. Each vector was subsequently normalized to unit length.

2. A blank image of size 512×512 pixels was created, and subsequently filled with randomly placed salt noise [36], that is, white pixels. The amount of salt noise was 5% of the total number of pixels in the image. The noisy image was blurred using a Gaussian filter of standard deviation 2 pixels, resulting in an image $I_{orig}(x, y)$.

3. An image with oriented texture was obtained by simulating the following differential equation over time:

$$\frac{\partial \rho(x, y, t)}{\partial t} + \nabla \cdot (\rho(x, y, t)\, \vec{v}(x, y)) = I_{orig}(x, y) ,$$

with the initial condition $\rho(x, y, 0) = 0$. This equation known as the continuity equation, or conservation-of-mass equation in fluid mechanics [47]. The noise image $I_{orig}(x, y)$ is interpreted as a constant source of matter, and the generated matter flows through space with the velocity $\vec{v}(x, y)$. The differential equation was simulated over the interval $t = 0$ to $t = 20$ (arbitrary units of time), and the final oriented texture image is given as $I(x, y) = \rho(x, y, 20)$.

The magnitude $M(x, y)$ and angle $\theta(x, y)$ components of the orientation field of $I(x, y)$ were obtained using the real Gabor filter bank described in Section 1.2.2. The filter bank is composed of 180 filters, spanning the range of orientations $0°$ to $179°$ in steps of $1°$. The parameters of each individual filter were $\tau = 4$ pixels and $l = 8$.

The textured image and the orientation field components were trimmed to a size of 256×256 pixels, by retaining only the central portion of each component. This step was performed in order to avoid border effects at the edges of $M(x, y)$ and $\theta(x, y)$. The resulting textured test images $I(x, y)$ and magnitude images $M(x, y)$ are shown in Figure 2.2.

The trimmed orientation field was analyzed using the procedure described in Section 2.2.2. The size of the analysis window was 10×10 pixels, which was slid one pixel per step. The phase portrait maps obtained were filtered with a Gaussian filter of standard deviation equal to 6 pixels, in order to aggregate votes that were in close proximity.

The resulting phase portrait maps can be observed in Figure 2.2. It is seen that there is a concentration of votes at the center of the phase portrait map corresponding to the type of phase portrait associated with each test image. It can also be noticed that votes have been cast in the phase portrait maps that do not correspond to the type of the phase portrait of each test image. Nevertheless, the miscast votes have been spread across the corresponding phase portrait map, and they do not result in the presence of a peak in the map that is as prominent as the peak present at the center of the correct phase portrait map.

2.3.2   ANALYSIS OF A MAMMOGRAPHIC ROI

The procedure for the analysis of large orientation fields, described in Section 2.2.2, was applied to a mammographic region of interest (ROI) exhibiting architectural distortion (from image mdb115 in the Mini-MIAS database). The analysis window size was 44×44 pixels, which was slid one pixel per step over the ROI of size 230×230 pixels. It can be observed in Figure 2.3 that the votes cast in the node map are less scattered than those cast in the saddle and spiral maps. Furthermore, the maximum value in the node map is larger than the maxima of the other two maps. This example illustrates that the phase portrait methodology can be useful in the detection of architectural distortion in mammograms.

Phase portrait type	Node	Saddle	Spiral
Matrix $\mathbf{A}$	$\begin{bmatrix} 1 & 0 \\ 0 & 2 \end{bmatrix}$	$\begin{bmatrix} 1 & 0 \\ 0 & -1 \end{bmatrix}$	$\begin{bmatrix} 1 & 1 \\ -1 & 1 \end{bmatrix}$
Test image	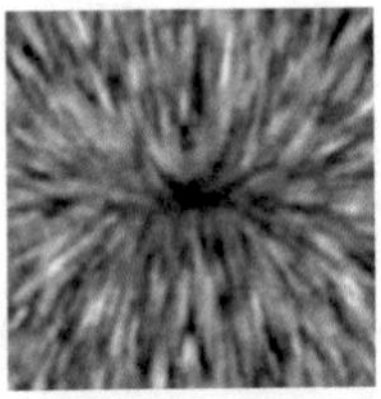	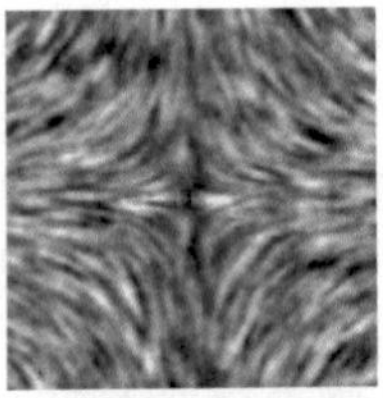	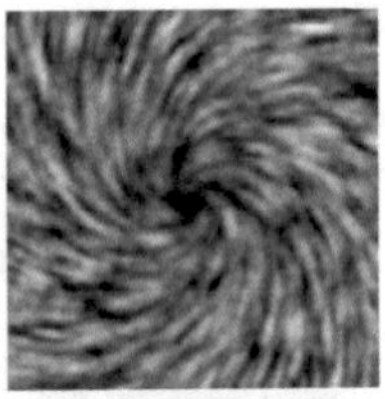
Magnitude field		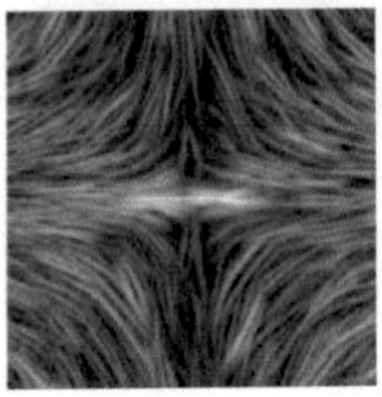	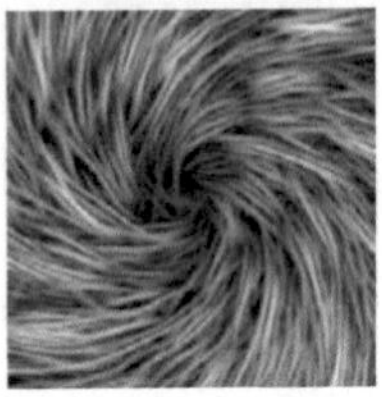

Figure 2.2 (cont.)

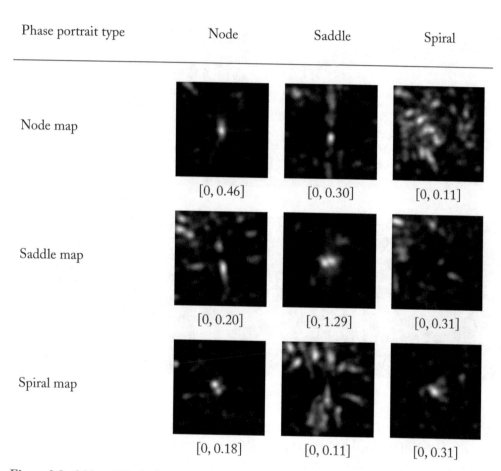

Phase portrait type	Node	Saddle	Spiral
Node map	[0, 0.46]	[0, 0.30]	[0, 0.11]
Saddle map	[0, 0.20]	[0, 1.29]	[0, 0.31]
Spiral map	[0, 0.18]	[0, 0.11]	[0, 0.31]

Figure 2.2: 256×256-pixel images corresponding to the experiment described in Section 2.3.1. *Matrix A*: the matrix **A** used in the generation of each test image. *Test image*: images displaying oriented texture. *Magnitude field*: the magnitude component of the orientation field of each test image. *Node, saddle, and spiral maps*: the corresponding phase portrait maps. The values shown below each map indicate its amplitude range.

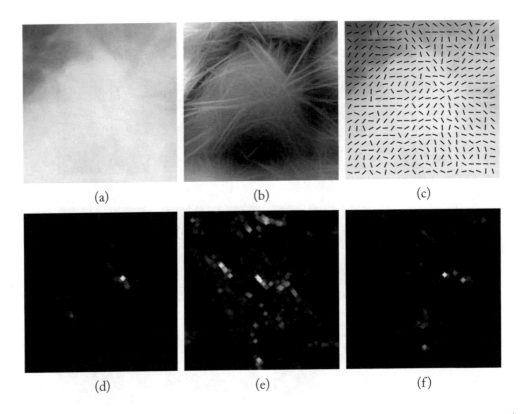

(a)   (b)   (c)

(d)   (e)   (f)

Figure 2.3: Analysis of an ROI from the image 'mdb115', which includes architectural distortion (see Figures 4.4 and 4.4): (a) ROI of size 230×230 pixels (46×46 mm); (b) magnitude image; (c) orientation field superimposed on the original ROI, with needles drawn for every 10th pixel; (d) node map, with intensities mapped from $[0, 123]$ to $[0, 255]$; (e) saddle map, $[0, 22]$ mapped to $[0, 255]$; (f) spiral map, $[0, 71]$ mapped to $[0, 255]$. The intensity values in each map correspond to the number of votes accumulated at each location. Reproduced with permission from F. J. Ayres and R. M. Rangayyan. "Characterization of architectural distortion in mammograms". IEEE Engineering in Medicine and Biology Magazine, 24(1):59–67, January 2005. © IEEE

2.4   REMARKS

This chapter presented the general methodology for the analysis of oriented texture using phase portraits, originally developed by Rao and Jain [44], as well as the application of the method to a set of test images and to a mammographic ROI displaying architectural distortion.

A study on optimization techniques for the fitting of the phase portrait model to a given orientation field is presented in Chapter 3. Methods for the detection of architectural distortion in full mammograms using phase portraits analysis are presented in Chapter 4.

CHAPTER 3

Optimization Techniques for Phase Portrait Analysis

3.1 INTRODUCTION

The analysis of oriented texture using phase portraits, as described in Chapter 2, requires the minimization of the error measure given in Equation 2.10 over the space of the model parameters. This is a nonlinear, nonconvex optimization problem, and practical experience shows that irrelevant local minima can lead to convergence to inappropriate results. As a consequence, the oriented texture under analysis may be incorrectly represented.

Rao and Jain [44, 45] used a nonlinear least-squares algorithm for the optimization of the model parameters. Other authors have investigated faster optimization procedures based on a linear approximation of the error measure [48, 49]. Algorithms for local-minimum search (such as nonlinear least-squares [50]) rely on an initial estimate of the optimal set of parameters, and improve upon the given estimate towards a local minimum of the objective function (i.e., the error measure). The dependence on initialization may render the local search algorithms inadequate for the application to phase portrait estimation.

In this chapter, a new optimization algorithm is presented, called the iterative linear least-squares algorithm [49, 51], for the accurate estimation of the phase portrait parameters, for a given orientation field. The performance of the new algorithm is investigated and compared with the performance of four optimization algorithms: linear least-squares [48], nonlinear least-squares [44], simulated annealing [52], and particle swarm optimization [53]. The algorithms are evaluated and compared in terms of the error between the estimated parameters and the parameters known by design, in the presence of noise in the orientation field and imprecision in the initialization of the parameters. The computational effort required by each algorithm is also assessed.

3.2 THE OPTIMIZATION PROBLEM

In this section, we analyze the error measure $\epsilon^2(\mathbf{A}, \mathbf{b})$, given in Equation 2.10. The analysis presented here will serve as the basis for the development of a new optimization procedure for the minimization of the error measure. Consider the phase portrait model presented in Equation 2.6, and reproduced here for convenience:

$$\vec{v} = \begin{pmatrix} v_x \\ v_y \end{pmatrix} = \mathbf{A} \begin{pmatrix} x \\ y \end{pmatrix} + \mathbf{b},$$

where

$$\mathbf{A} = \begin{bmatrix} a & b \\ c & d \end{bmatrix}, \quad \mathbf{b} = \begin{bmatrix} e \\ f \end{bmatrix}.$$

The orientation field generated by the phase portrait model is defined in Equation 2.8, and reproduced in the following:

$$\phi(x, y | \mathbf{A}, \mathbf{b}) = \arctan\left(\frac{v_y}{v_x}\right).$$

From Equation 2.6, we have

$$\sin(\phi(x, y | \mathbf{A}, \mathbf{b})) = \frac{v_y}{\sqrt{v_x^2 + v_y^2}} = \frac{cx + dy + f}{\sqrt{v_x^2 + v_y^2}}, \tag{3.1}$$

and

$$\cos(\phi(x, y | \mathbf{A}, \mathbf{b})) = \frac{v_x}{\sqrt{v_x^2 + v_y^2}} = \frac{ax + by + e}{\sqrt{v_x^2 + v_y^2}}. \tag{3.2}$$

From Equations 3.1 and 3.2, it follows that multiplying the parameters of the phase portrait model (the matrix $\mathbf{A}$ and the vector $\mathbf{b}$) by a constant does not change the orientation field generated by the phase portrait model.

Let us define x_i and y_i as the x and y coordinates of the ith pixel, $1 \leq i \leq N$. Let us also define $\theta_i = \theta(x_i, y_i)$, and $\phi_i = \phi(x_i, y_i | \mathbf{A}, \mathbf{b})$. As discussed in Chapter 2, the error between the angle of the phase portrait obtained from the model and the angle derived from the image at the ith pixel, named the local error measure, is defined as

$$r_i = \sin(\theta_i - \phi_i), \tag{3.3}$$

which could be expanded as

$$r_i = \sin(\theta_i) \cos(\phi_i) - \cos(\theta_i) \sin(\phi_i). \tag{3.4}$$

Using Equations 3.1, 3.2, and 3.4, we obtain

$$r_i = \frac{(x_i \sin\theta_i)a + (y_i \sin\theta_i)b + (\sin\theta_i)e}{\sqrt{(v_x)_i^2 + (v_y)_i^2}} + \frac{(-x_i \cos\theta_i)c + (-y_i \cos\theta_i)d + (-\cos\theta_i)f}{\sqrt{(v_x)_i^2 + (v_y)_i^2}}. \tag{3.5}$$

The error vector is given by $\mathbf{r} = [r_1, r_2, \cdots, r_N]^T$, and the sum of the squared error is given by

$$\epsilon^2(\mathbf{A}, \mathbf{b}) = \mathbf{r}^T \mathbf{r} = \sum_{i=1}^{N} r_i^2 \,. \tag{3.6}$$

Minimization of $\epsilon^2(\mathbf{A}, \mathbf{b})$ leads to the optimal phase portrait parameters that describe the orientation field of the image under analysis.

3.3   OPTIMIZATION PROCEDURES

3.3.1   LINEAR LEAST-SQUARES

Shu and Jain [48] proposed the use of a linear least-squares algorithm to find the optimal phase portrait parameters, using an orientation field error formulation that is different from that given in Equation 3.6. The following derivation developed by us [49] differs from the derivation given by Shu and Jain, while arriving at the same final equations for the linear least-squares procedure. The present derivation leads to the iterative linear least-squares method for the optimization of phase portrait parameters [49], which is presented in Section 3.3.2.

Let us define a modified error measure at the ith pixel as $s_i = r_i \sqrt{(v_x)_i^2 + (v_y)_i^2}$. Then, from Equation 3.5, we obtain

$$s_i = (x_i \sin \theta_i)a + (y_i \sin \theta_i)b + (-x_i \cos \theta_i)c + (-y_i \cos \theta_i)d + (\sin \theta_i)e + (-\cos \theta_i)f, \tag{3.7}$$

which is a linear function of the phase portrait parameters. The modified error vector is defined as $\mathbf{s} = [s_1, s_2, \cdots, s_N]^T$, and the sum of the squared error for the orientation field is

$$\epsilon_l^2(\mathbf{A}, \mathbf{b}) = \mathbf{s}^T \mathbf{s} \,. \tag{3.8}$$

A trivial solution to the problem of minimizing $\epsilon_l^2(\mathbf{A}, \mathbf{b})$ is to set all of the parameters $[a, b, c, d, e, f]$ to zero; in order to avoid the trivial solution, the constraint $a^2 + b^2 + c^2 + d^2 = 1$ is imposed. A linear least-squares algorithm can be used to estimate the optimal parameters of the phase portrait model in order to minimize $\epsilon_l^2(\mathbf{A}, \mathbf{b})$, with the formulation as follows.

Define the parameter vectors $\mathbf{u} = [a, b, c, d]^T$ and $\mathbf{v} = [e, f]^T$, and the matrices

$$\mathbf{U} = \begin{bmatrix} x_1 \sin \theta_1 & y_1 \sin \theta_1 & -x_1 \cos \theta_1 & -y_1 \cos \theta_1 \\ x_2 \sin \theta_2 & y_2 \sin \theta_2 & -x_2 \cos \theta_2 & -y_2 \cos \theta_2 \\ \vdots & \vdots & \vdots & \vdots \\ x_N \sin \theta_N & y_N \sin \theta_N & -x_N \cos \theta_N & -y_N \cos \theta_N \end{bmatrix}, \quad \mathbf{V} = \begin{bmatrix} \sin \theta_1 & -\cos \theta_1 \\ \sin \theta_2 & -\cos \theta_2 \\ \vdots & \vdots \\ \sin \theta_N & -\cos \theta_N \end{bmatrix}.$$

Then, the error vector is given as

$$s = Uu + Vv. \tag{3.9}$$

Imposing the constraint $u^T u = a^2 + b^2 + c^2 + d^2 = 1$, the constrained optimization problem is to minimize the cost function

$$P = s^T s + \lambda(u^T u - 1) = (Uu + Vv)^T(Uu + Vv) + \lambda(u^T u - 1), \tag{3.10}$$

where λ is the Lagrange multiplier. Differentiating P with respect to u, v, and λ yields

$$\frac{\partial P}{\partial u} = 2U^T Uu + 2U^T Vv + 2\lambda u = 0,$$

$$\frac{\partial P}{\partial v} = 2V^T Vv + 2V^T Uu = 0,$$

$$\frac{\partial P}{\partial \lambda} = u^T u - 1 = 0.$$

Solving the system of equations above for the variables u and v, we obtain

$$u^T u = 1, \tag{3.11}$$

$$v = -(V^T V)^{-1} V^T Uu, \tag{3.12}$$

$$\Psi u = \lambda u, \tag{3.13}$$

where $\Psi = U^T V(V^T V)^{-1} V^T U - U^T U$. Substituting the above equations into the cost function, we obtain $P = -\lambda$. Therefore, the optimal u vector is the eigenvector of Ψ associated with the largest eigenvalue. Using Equation 3.12, the solution for v is obtained from the optimal value of u.

3.3.2   ITERATIVE LINEAR LEAST-SQUARES

Observe that the linear least-squares procedure minimizes the modified error measure $\epsilon_l^2(A, b)$ in Equation 3.8, instead of the original error measure $\epsilon^2(A, b)$ in Equation 3.6. We propose an iterative, weighted estimation algorithm [49] to achieve improved minimization of the original error $\epsilon^2(A, b)$, based on the linear least-squares algorithm presented above. Let D_j be an $N \times N$ diagonal matrix, which we shall call as the *correction matrix*, with the diagonal elements defined as

$$d_{i,i} = \frac{1}{\sqrt{(v_x)_i^2 + (v_y)_i^2}}. \tag{3.14}$$

Let M be the number of iterations (predetermined empirically as $M = 5$). The algorithm proceeds as follows:

1. Set $\mathbf{D}_0 = \mathbf{I}$, the identity matrix.

2. For $j = 1$ to M

    (a) Apply the correction matrix to the data matrices: replace $\mathbf{U}$ by $\mathbf{D}_{j-1}\mathbf{U}$, and $\mathbf{V}$ by $\mathbf{D}_{j-1}\mathbf{V}$.

    (b) Estimate the parameters $[a, b, c, d, e, f]$.

    (c) Evaluate $\mathbf{D}_j$ with the new estimated parameters.

Notice that the iterative algorithm reduces to that of Shu and Jain [48] for $M = 1$. The computational cost of each iteration is small, compared to the use of the nonlinear least-squares method as proposed by Rao and Jain [44]. As a consequence, it is possible to iterate the algorithm in a practical situation without excessive computational burden.

At each iteration, it is expected that the algorithm will result in an improved estimate of the true parameters of the phase portrait model. However, due to the fact that the algorithm requires a correction factor that is inversely proportional to the magnitude of $\vec{v}(x, y)$, numerical instabilities are possible. Hence, the algorithm is limited to a fixed number of iterations. Potential improvements to the procedure include the incorporation of a regularization term to avoid numerical instabilities.

3.3.3   NONLINEAR LEAST-SQUARES

In their work on the analysis of oriented texture using phase portraits, Rao and Jain [44] used a nonlinear least-squares algorithm for the minimization of the error measure. The nonlinear least-squares algorithm is a local-minimum search procedure where the nonlinear function to be minimized is formed by the sum of squared functions. A general formulation of the nonlinear least-squares procedure is as follows. Let $\epsilon(\mathbf{z})$ be a function to be minimized of the vector of variables $\mathbf{z}$, given by the sum of squared functions $f_i(\mathbf{z})$ as follows:

$$\epsilon(\mathbf{z}) = \frac{1}{2} \sum_{i=1}^{N} f_i(\mathbf{z}) = \frac{1}{2}||\mathbf{f}(\mathbf{z})||^2 \, ,$$

where $\mathbf{f}(\mathbf{z}) = [f_1(\mathbf{z}), f_2(\mathbf{z}), \cdots, f_N(\mathbf{z})]^T$, and $||.||$ denotes the L^2-norm of the vector. All nonlinear least-squares methods proceed from an initial guess $\mathbf{z}_0$ of the optimal parameters, and at each iteration, the following linearized optimization problem is solved:

$$\text{minimize} \quad \eta(\Delta\mathbf{z}) = \frac{1}{2}||\mathbf{f}(\mathbf{z}_i) + \mathbf{J}\Delta\mathbf{z}|| \, ,$$

where $\mathbf{J}$ is the Jacobian matrix of $\mathbf{f}(\mathbf{z})$ computed at $\mathbf{z}_i$, and $\Delta\mathbf{z}$ is a step in the parameter space towards an improved estimate. The next estimate of the optimal parameters is given by $\mathbf{z}_{i+1} = \mathbf{z}_i + \alpha\Delta\mathbf{z}$, where α is the step size. The parameter α is computed differently in each variation of the nonlinear least-squares algorithm, and it represents a balance between faster convergence (greater α) and higher stability (smaller α).

In the context of the estimation of the phase portrait parameters, $\mathbf{f}(\mathbf{z}) = \mathbf{r}$, where $\mathbf{z} = [a, b, c, d, e, f]^T$. The nonlinear least-squares algorithm used in this investigation is the Levenberg-Marquardt method [50], implemented in the GNU Scientific Library [54], a collection of routines for numerical computation.

3.3.4   SIMULATED ANNEALING

The simulated annealing algorithm [52] is a global optimization procedure. The algorithm is iterative, and at each iteration, a new estimate of the optimal parameters is computed by taking a random step of limited length in the parameter space, from the current parameter vector. The new parameter vector is accepted as a new estimate of the optimal parameters if it represents an improved solution (the value of the function to be minimized is smaller at the new parameter vector than that at the current parameter vector). If the new parameters correspond to a worse solution, the parameters may be accepted with a probability that decreases as the computation proceeds. If the probability of accepting a worse solution is decreased at a slow rate as the iterations progress, the algorithm will converge to the global optimum of the function being optimized. The algorithm is presented in Figure 3.1 in pseudo-code for greater clarity.

3.3.5   PARTICLE SWARM OPTIMIZATION

The particle swarm optimization algorithm constructs an analogy between the behavior of a swarm of insects, searching for the optimal site on a feeding ground, and the search for the optimal point of a function. The algorithm was developed by Kennedy and Eberhart [53] (see also Clerc and Kennedy [55]), and it has been shown to be successful in various applications where global optimization is required. Figure 3.2 gives a pseudo-code description of the particle swarm optimization algorithm.

3.4   COMPARATIVE ANALYSIS OF THE OPTIMIZATION PROCEDURES

The performance of the four optimization procedures described in Section 3.3 was investigated using three orientation fields of size 41×41 pixels, generated according to the parameters in Table 3.1. Uniformly distributed noise was added to the orientation field angles. The phase portrait parameters were estimated from the noisy orientation fields, for several levels of noise, using each optimization procedure. Four combined approaches were also investigated for the estimation of the phase portrait parameters as follows:

- Each of the linear least-squares, iterative linear least-squares, simulated annealing, and particle swarm optimization procedures was applied to find an initial estimate of the optimal phase portrait parameters.

Simulated Annealing algorithm
Input:

- a function $f(\mathbf{z})$ to be minimized over the parameter space of the vector $\mathbf{z}$
- R: cooling rate
- $\mathbf{z}_0$: an initial estimate of the optimal parameters
- N: number of cooling cycles
- T: initial temperature
- randomstep(): a function that returns a random vector of limited length in the parameter space
- rnd(): a function that returns a random real number in the interval $[0, 1]$

Output

- $\mathbf{z}_{opt}$: final estimate of the optimal parameters of $f(\mathbf{z})$

Algorithm

$\mathbf{z}_{opt} \leftarrow \mathbf{z}_0$

for k = 1 to N

    $\mathbf{z}_{new} \leftarrow \mathbf{z}_{opt} + \text{randomstep}()$
    if $f(\mathbf{z}_{new}) < f(\mathbf{z}_{opt})$

        $\mathbf{z}_{opt} \leftarrow \mathbf{z}_{new}$

    else

        if $\text{rnd}() < \exp\left(-\dfrac{f(\mathbf{z}_{new}) - f(\mathbf{z}_{opt})}{T R^k}\right)$
          $\mathbf{z}_{opt} \leftarrow \mathbf{z}_{new}$
        end

    end

end

Figure 3.1: Pseudo-code description of the simulated annealing algorithm.

Particle Swarm Optimization algorithm
Input:

- a function $f(\mathbf{z})$ to be minimized over the parameter space of the vector $\mathbf{z}$

- N_{iter}: number of iterations

- N_{part}: number of particles

- ϕ_1, ϕ_2, χ: parameters of the algorithm [55]

- rnd(): a function that returns a random real number in the interval [0, 1]

Output

- $\mathbf{z}_{opt}$: final estimate of the optimal parameters of $f(\mathbf{z})$

Algorithm

Initialize randomly positions $\mathbf{x}_i$ and velocities $\mathbf{v}_i$ for each particle. Set best visited position of each particle as $\mathbf{p}_i = \mathbf{x}_i$. Set $\mathbf{p}_{best}$ to the value of $\mathbf{p}_i$ that minimizes $f(\mathbf{p}_i)$ over the set of particles.

for k = 1 to N_{iter}

- Update the velocity of each particle as
$$\mathbf{v}_i \leftarrow \chi \left(\mathbf{v}_i + \phi_1 \left(\mathbf{p}_i - \mathbf{x}_i\right) \mathrm{rnd}() + \phi_2 \left(\mathbf{p}_{best} - \mathbf{x}_i\right) \mathrm{rnd}()\right)$$
- Update the position of each particle as $\mathbf{x}_i = \mathbf{x}_i + \mathbf{v}_i$
- Update $\mathbf{p}_i$ for each particle
- Update $\mathbf{p}_{best}$

end

$\mathbf{z}_{opt} = \mathbf{p}_{best}$

Figure 3.2: Pseudo-code description of the particle swarm optimization algorithm algorithm. The parameters of the algorithm were chosen as $\phi_1 = 2.05$, $\phi_2 = 2.05$, and $\chi = 0.8543$, as recommended by Clerc and Kennedy [55]. The number of particles was chosen empirically as 500, and the number of iterations as 40.

- The nonlinear least-squares procedure was used to improve further the estimate of the phase portrait parameters from the preceding step.

The nonlinear least-squares (except in the combined approaches) and the simulated annealing algorithms were initialized with the following values:

$$\mathbf{A}_0 = \begin{bmatrix} \sqrt{2} & 0 \\ 0 & \sqrt{2} \end{bmatrix} \text{ and } \mathbf{b}_0 = \begin{bmatrix} 0 \\ 0 \end{bmatrix}.$$

The performance of the optimization algorithms was measured in terms of the error in the estimated parameters, as well as the distance between the estimated fixed-point location and the true fixed-point location (fixed-point error, in pixels). The computational time required by each optimization procedure was also recorded.

The parameter error is defined as follows. Let $\mathbf{p}$ be the true parameter vector, and $\mathbf{q}$ be the estimated parameter vector. The error between the parameter vectors is defined as

$$\text{Parameter error} = 1 - \frac{|<\mathbf{p},\mathbf{q}>|}{\|\mathbf{p}\|\|\mathbf{q}\|}, \tag{3.15}$$

where $<\mathbf{p},\mathbf{q}>$ is the dot product of $\mathbf{p}$ and $\mathbf{q}$.

The parameter error, fixed-point error, and computational time were measured for each noise level and test pattern. Each experiment was repeated 100 times, and the results averaged. The range of noise was varied from $[-5, 5]$ to $[-45, 45]$ degrees. The noise amplitude varied from 0 to 45 degrees.

Figures 3.3, 3.4, and 3.5 illustrate the results obtained for each test pattern in one set of experiments. The computational time required by each method is shown in Table 3.2. The following observations were made:

- The simulated annealing method yielded low fixed-point and parameter errors in all tests, and the fixed-point error was almost the same for all noise levels.

- The linear least-squares and the iterative linear least-squares methods resulted in fixed-point and parameter errors that increased with the noise level. The iterative linear least-squares method outperformed the linear least-squares method in terms of fixed-point error. The two methods are comparable in terms of parameter error.

- The nonlinear least-squares algorithm performed differently between the tests. A good performance in both fixed-point and parameter errors was observed with the node and spiral patterns, but a poor performance was obtained with the saddle pattern.

- All of the combined approaches yielded the same performance in terms of fixed-point and parameter errors, with the exception of an increased error for the linear least-squares method at higher noise levels with the spiral pattern. The performance of the combined methods was superior to those of the corresponding individual optimization procedures.

Table 3.1: Parameters of the orientation field generated for each test. The location of the fixed point is $\mathbf{x}_0$. Reproduced with permission from F. J. Ayres and R. M. Rangayyan. "Optimization procedures for the estimation of phase portrait parameters of orientation fields". In E. R. Dougherty, J. T. Astola, K. O. Egiazarian, N. M. Nasrabadi, and S. A. Rizvi, editors, Proceedings of SPIE Electronic Imaging 2006, volume 6064, San Jose, CA, 2006. © SPIE

Test #	A	$\mathbf{x}_0$	Type
1	$\begin{bmatrix} 1 & 0 \\ 0 & 1.5 \end{bmatrix}$	$\begin{bmatrix} 10 \\ 10 \end{bmatrix}$	node
2	$\begin{bmatrix} 1 & 0 \\ 0 & -1.5 \end{bmatrix}$	$\begin{bmatrix} 10 \\ 10 \end{bmatrix}$	saddle
3	$\begin{bmatrix} 1 & -1 \\ 1 & 1 \end{bmatrix}$	$\begin{bmatrix} 10 \\ 10 \end{bmatrix}$	spiral

- The linear and the iterative linear least-squares methods are the fastest procedures. The simulated annealing procedure is the slowest method, with a running time one order of magnitude higher than that of the nonlinear least-squares and the combined approaches involving the linear and the iterative linear least-squares methods.

Table 3.2: Execution time for the optimization methods. LIN: Linear least-squares. ILIN: Iterative linear least-squares. SA: Simulated annealing. PSO: Particle swarm optimization. NLS: Nonlinear least-squares. LIN + NLS: combined method using the linear least-squares followed by the nonlinear least-squares method. ILIN + NLS: combined method using the iterative linear least-squares followed by the nonlinear least-squares method. SA + NLS: combined method using the simulated annealing followed by the nonlinear least-squares method. PSO + NLS: combined method using the particle swarm optimization followed by the nonlinear least-squares method. The computer used is a Dell Precision 360 workstation, with an Intel Pentium IV 3.2 GHz CPU (2MB cache) and 2 GB of RAM. Reproduced with permission from F. J. Ayres and R. M. Rangayyan. "Optimization procedures for the estimation of phase portrait parameters of orientation fields". In E. R. Dougherty, J. T. Astola, K. O. Egiazarian, N. M. Nasrabadi, and S. A. Rizvi, editors, Proceedings of SPIE Electronic Imaging 2006, volume 6064, San Jose, CA, 2006. © SPIE

Method	Test pattern		
	node	saddle	spiral
LIN	< 1 ms	< 1 ms	< 1 ms
ILIN	< 1 ms	< 1 ms	< 1 ms
SA	0.11 s	0.11 s	0.11 s
PSO	1.05 s	1.05 s	1.05 s
NLS	0.03 s	0.28 s	0.04 s
LIN + NLS	0.02 s	0.02 s	0.08 s
ILIN + NLS	0.02 s	0.02 s	0.06 s
SA + NLS	0.13 s	0.13 s	0.17 s
PSO + NLS	1.08 s	1.08 s	1.10 s

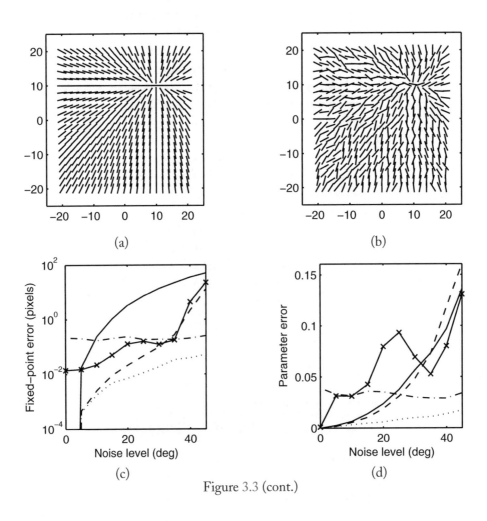

Figure 3.3 (cont.)

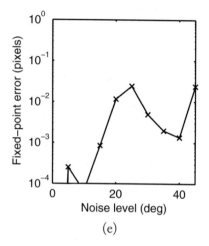

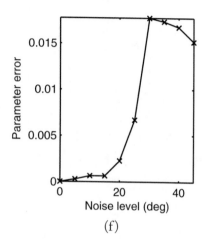

(e)

(f)

Figure 3.3: Results of the experiment with the node test pattern. (a) The original orientation field with 41×41 pixels. (b) The orientation field corrupted by uniformly distributed angle noise in the range $[-30, 30]$ degrees. (c) The fixed-point error for each optimization procedure. (d) The parameter error for each optimization procedure. (e) The fixed-point error for the combined procedures (as listed in Table 3.2). (f) The parameter error for the combined procedures. For (e) and (f): the nonlinear least-squares method was applied in combination with each of the remaining four procedures. The four plots in (e) and (f) overlap completely. (Solid line: linear least-squares; dashed line: iterative linear least-squares; dash-dot line: simulated annealing; solid-with-crosses: particle swarm optimization; dotted line: nonlinear least-squares. deg: degrees) Reproduced with permission from F. J. Ayres and R. M. Rangayyan. "Optimization procedures for the estimation of phase portrait parameters of orientation fields". In E. R. Dougherty, J. T. Astola, K. O. Egiazarian, N. M. Nasrabadi, and S. A. Rizvi, editors, Proceedings of SPIE Electronic Imaging 2006, volume 6064, San Jose, CA, 2006. © SPIE

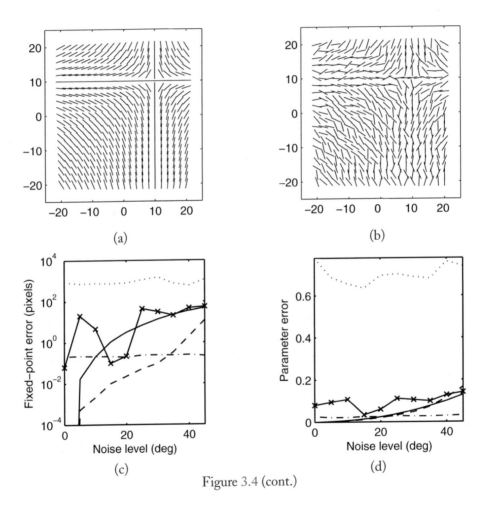

(a)

(b)

(c)

(d)

Figure 3.4 (cont.)

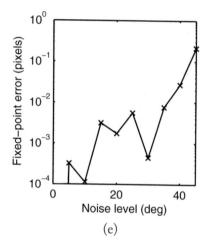

(e)

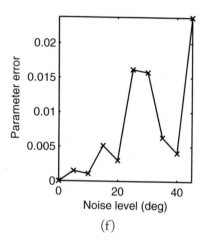

(f)

Figure 3.4: Results of the experiment with the saddle test pattern. (a) The original orientation field with 41 × 41 pixels. (b) The orientation field corrupted by uniformly distributed angle noise in the range [−30, 30] degrees. (c) The fixed-point error for each optimization procedure. (d) The parameter error for each optimization procedure. (e) The fixed-point error for the combined procedures (as listed in Table 3.2). (f) The parameter error for the combined procedures. For (e) and (f): the nonlinear least-squares method was applied in combination with each of the remaining four procedures. The four plots in (e) and (f) overlap completely. (Solid line: linear least-squares; dashed line: iterative linear least-squares; dash-dot line: simulated annealing; solid-with-crosses: particle swarm optimization; dotted line: nonlinear least-squares. deg: degrees). Reproduced with permission from F. J. Ayres and R. M. Rangayyan. "Optimization procedures for the estimation of phase portrait parameters of orientation fields". In E. R. Dougherty, J. T. Astola, K. O. Egiazarian, N. M. Nasrabadi, and S. A. Rizvi, editors, Proceedings of SPIE Electronic Imaging 2006, volume 6064, San Jose, CA, 2006. © SPIE

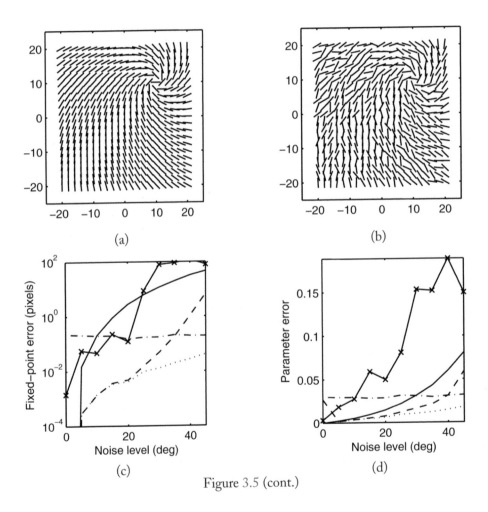

(a)

(b)

(c)

(d)

Figure 3.5 (cont.)

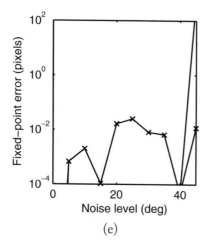

(e)

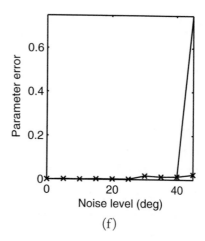

(f)

Figure 3.5: Results of the experiment with the spiral test pattern. (a) The original orientation field with 41 × 41 pixels. (b) The orientation field corrupted by uniformly distributed angle noise in the range [−30, 30] degrees. (c) The fixed-point error for each optimization procedure. (d) The parameter error for each optimization procedure. (e) The fixed-point error for the combined procedures (as listed in Table 3.2). (f) The parameter error for the combined procedures. For (e) and (f): the nonlinear least-squares method was applied in combination with each of the remaining four procedures. The four plots in (e) and (f) overlap significantly. (Solid line: linear least-squares; dashed line: iterative linear least-squares; dash-dot line: simulated annealing; solid-with-crosses: particle swarm optimization; dotted line: nonlinear least-squares. deg: degrees). Reproduced with permission from F. J. Ayres and R. M. Rangayyan. "Optimization procedures for the estimation of phase portrait parameters of orientation fields". In E. R. Dougherty, J. T. Astola, K. O. Egiazarian, N. M. Nasrabadi, and S. A. Rizvi, editors, Proceedings of SPIE Electronic Imaging 2006, volume 6064, San Jose, CA, 2006. © SPIE

3.5 REMARKS

This chapter presented a new optimization method (iterative linear least-squares) for the estimation of the optimal phase portrait parameters for a given orientation field, and the results of an investigation on the performance of the new method and four other optimization algorithms: linear least-squares, nonlinear least-squares, simulated annealing, and particle swarm optimization. The results lead to the following observations:

- The application of the nonlinear least-squares algorithm to the problem of phase portrait estimation leads to results that are sensitive to the initial estimate (initialization) of the model parameters. The initial parameters of the algorithm in the examples presented correspond to a node pattern, which explains the good performance of the nonlinear least-squares algorithm in the node and spiral cases. In the case of the spiral pattern, the phase portrait converges to a central point in a manner that is similar to the case with the node test pattern.

- The linear least-squares algorithm does not require initialization; nevertheless, the algorithm requires a linear approximation of the error measure, which reduced the precision of the estimated parameters in some instances.

- The combined approaches resulted in similar performance, with generally improved solutions as compared to those obtained with the application of the corresponding individual optimization procedures. The combination of the simulated annealing and the nonlinear least-squares methods yielded the best performance among all of the methods tested.

- The simulated annealing algorithm is more robust to variations in initialization than the nonlinear least-squares method, while being computationally more expensive than all of the other methods tested except the particle swarm optimization method.

- The expected computational time requirement of the iterative linear least-squares method is higher than the time requirement of the linear least-squares method. In the experiments illustrated in this chapter, both methods were faster than the time resolution adopted (1 ms). The observation that the iterative linear least-squares method yielded a higher accuracy in the estimation of the phase portrait parameters than the linear least-squares method leads to the conclusion that the adoption of the former is preferred to the latter.

It is concluded that the application of simulated annealing followed by the nonlinear least-squares method is the preferred optimization strategy in applications where a high level of noise is present. When the oriented texture in the given image is clearly evident and highly coherent (the orientation field has a local dominant orientation at every pixel), a faster optimization procedure may be used, such as the combination of the iterative linear least-squares and the nonlinear least-squares methods.

CHAPTER 4

Detection of Sites of Architectural Distortion in Mammograms

Among the most commonly missed signs of breast cancer is architectural distortion, which is defined in the Breast Imaging Reporting and Data System (BI-RADS) [84] as follows: "The normal architecture is distorted with no definite mass visible. This includes spiculations radiating from a point and focal retraction or distortion at the edge of the parenchyma." According to Homer [76]: "Architectural distortion is a localizing sign of breast cancer produced by a desmoplastic reaction; its presence demands an explanation. Some benign etiologies for this finding, such as previous biopsy and inflammation can be suspected by history. If no explanation for the architectural distortion can be elicited, biopsy is often the next indicated procedure."

A normal mammogram is shown in Figure 4.1, and a region of interest (ROI) of the normal mammogram is presented in Figure 4.2. It is seen that the oriented texture in the normal mammogram is coherently aligned. Figure 4.3 shows a mammogram presenting architectural distortion, made evident by the presence of radial spicules. In Figure 4.4, the region of architectural distortion of the preceding mammogram is shown in enlarged detail. Observe that the density of the fibroglandular disk partially obscures the visibility of the spicules.

Sickles [152] reported that indirect signs of malignancy (such as architectural distortion, bilateral asymmetry, single dilated duct, and developing densities) accounted for almost 20% of the detected cancers. Burrell *et al.* [153] observed that architectural distortion was the most commonly missed abnormality in false-negative cases, in a study of cases of screening interval breast cancer. Architectural distortion is a subtle abnormality, and improvements in the detection rate of architectural distortion could increase the rate of detection of early breast cancer, and reduce the morbidity and mortality due to breast cancer. In this chapter, we present methods for the detection of architectural distortion in full mammograms.

4.1 THE FIRST METHOD: BASIC ANALYSIS OF THE NODE MAP

In a preliminary study [56] on the detection of architectural distortion in mammograms, a method was developed comprising the following steps (see the flowchart in Figure 4.5):

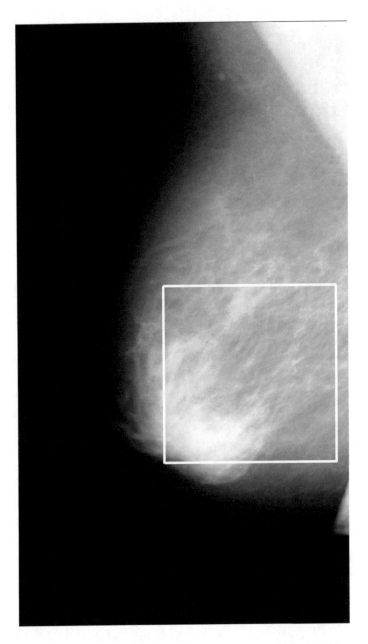

Figure 4.1: Normal mammogram from the Mini-MIAS database (mdb071). True width of image = 580 pixels = 116 mm. The square box overlaid on the figure represents the ROI, shown enlarged in Figure 4.2.

Figure 4.2: Detail of a normal mammogram (mdb071) showing the ROI marked by the box in Figure 4.1. True width of image = 300 pixels = 60 mm.

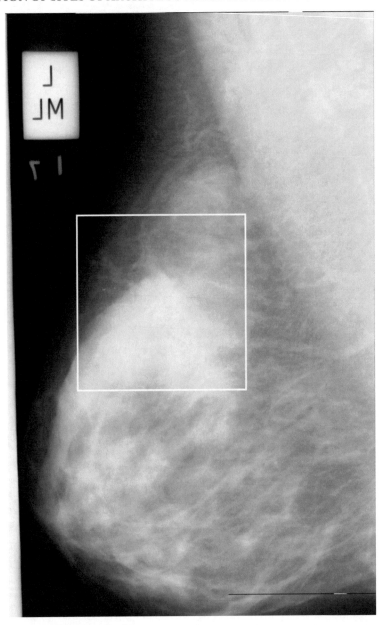

Figure 4.3: Architectural distortion present in a mammogram from the Mini-MIAS database (mdb115). True width of image = 650 pixels = 130 mm. The square box overlaid on the figure represents the ROI including the site of architectural distortion, shown enlarged in Figure 4.4. Reproduced with permission from F. J. Ayres and R. M. Rangayyan. "Characterization of architectural distortion in mammograms". IEEE Engineering in Medicine and Biology Magazine, 24(1):59–67, January 2005. © IEEE

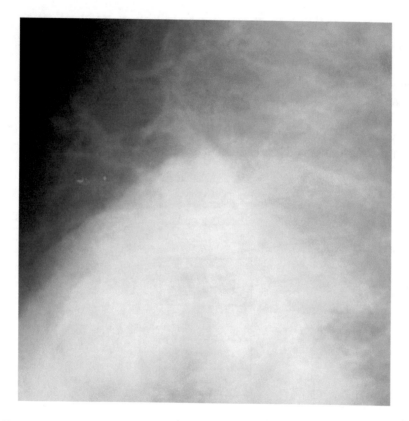

Figure 4.4: Detail of mammogram mdb115 showing the site of architectural distortion marked by the box in Figure 4.3. True width of image = 300 pixels = 60 mm. Reproduced with permission from F. J. Ayres and R. M. Rangayyan. "Characterization of architectural distortion in mammograms". IEEE Engineering in Medicine and Biology Magazine, 24(1):59–67, January 2005. © IEEE

- Computation of the orientation field using real Gabor filters, as described in Section 1.2.2. The parameters of the Gabor filters are $\tau = 4$ pixels, $l = 8$, and the filter bank consists of $K = 180$ filters, spanning the range $[-\pi/2, \pi/2]$ radians.

- Filtering and downsampling of the orientation field, described in Section 4.1.1.

- Phase portrait modeling of the filtered and downsampled orientation field, as described in Section 2.2.2. The size of the analysis window employed is 10×10 pixels.

- Post-processing of the node map and detection of sites of architectural distortion, as described in Section 4.1.2.

This method will be referred to as the "first method" in the rest of the chapter. An evaluation of the performance of this method is presented in Section 4.1.3.

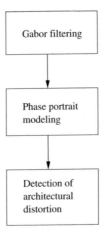

Figure 4.5: Flowchart for the first method for the detection of architectural distortion.

4.1.1 FILTERING AND DOWNSAMPLING THE ORIENTATION FIELD

The orientation field needs to be filtered and downsampled in order to reduce noise and also to reduce the computational effort required for the processing of full mammograms. Let $h(x, y)$ be a Gaussian filter of standard deviation σ_f, defined as

$$h(x, y) = \frac{1}{2\pi\sigma_f} \exp\left[-\frac{1}{2}\left(\frac{x^2 + y^2}{\sigma_f^2}\right)\right].$$

(4.1)

Define the images $s(x, y) = \sin[2\theta(x, y)]$ and $c(x, y) = \cos[2\theta(x, y)]$, where $\theta(x, y)$ is the orientation field angle. Then, the filtered orientation field angle $\theta_f(x, y)$ is obtained as

$$\theta_f(x, y) = \frac{1}{2} \arctan\left(\frac{(h * s)(x, y)}{(h * c)(x, y)}\right), \tag{4.2}$$

where the asterisk denotes convolution.

The filtered orientation field is downsampled by a factor of four, thus producing the downsampled orientation field θ_d as

$$\theta_d(x, y) = \theta_f(4x, 4y). \tag{4.3}$$

The filtering procedure employed in this work is a variant of Rao's dominant local orientation method [45]: in our procedure, we use a Gaussian filter instead of a box filter.

4.1.2 POST-PROCESSING OF THE NODE MAP AND DETECTION IN THE FIRST METHOD

We use a simple procedure to detect and locate sites of architectural distortion, using only the node map, as follows:

1. The node map is filtered with a Gaussian filter of standard deviation equal to 1.0 pixel (0.8 mm).

2. The filtered node map is thresholded (the threshold value is the same for all images).

3. The thresholded image is subjected to the following series of morphological operations [57] to group positive responses that are close to one another and to reduce each region of positive response to a single point. The resulting points indicate the detected locations of architectural distortion.

   (a) A closing operation is performed to group clusters of points that are less than 8 mm apart. The structural element is a disk of radius 10 pixels (8 mm).

   (b) A "close holes" filter is applied to the image. The resulting image includes only compact regions.

   (c) The image is subjected to a "shrink" filter where each compact region is shrunk to a single pixel.

The threshold value influences the sensitivity of the method and the resulting number of false positives per image. A high threshold value reduces the number of false positives, but it also reduces the sensitivity. A low threshold value increases the number of false positives.

4.1.3   RESULTS OF THE FIRST METHOD

The proposed method was applied to 18 mammograms exhibiting architectural distortion, selected from the Mini-MIAS database [42]. The mammograms are medio-lateral oblique (MLO) views, digitized to 1024×1024 pixels at a resolution of 200 μm and 8 bits/pixel. Figures 4.6 and 4.7 illustrate the steps of the method, as applied to image 'mdb115'. Observe that the filtered orientation field (Figure 4.6d) is smoother and more coherent as compared to the original orientation field (Figure 4.6c): the pattern of architectural distortion is better displayed in the filtered orientation field.

The architectural distortion present in the mammogram 'mdb115' has a stellate or spiculated appearance. As a consequence, a large number of votes have been cast into the node map, at a location close to the center of the distortion, as seen in Figure 4.7a. Another point of accumulation of votes in the node map is observed in Figure 4.7a, at the location of the nipple. This is not unexpected: the breast has a set of lactiferous ducts that convey milk to the nipple; the ducts appear in mammograms as linear structures converging to the nipple. Observe that the node map has the strongest response of all maps, within the site of architectural distortion given by the Mini-MIAS database. Our technique has resulted in the identification of two locations as sites of architectural distortion: one true positive and one false positive, as shown in Figure 4.7d.

The free-response receiver operating characteristics (FROC) were derived by varying the threshold level in the detection step; the result is shown in Figure 4.8. A sensitivity of 88% was obtained at 15 false positives per image.

4.2   THE SECOND METHOD: SELECTION OF CURVILINEAR STRUCTURES AND NODE EMPHASIS

With the aim of reducing the number of false positives, the second method [14, 58] incorporates the following modifications with respect to the first method presented in Section 4.1 (see the flowchart in Figure 4.9):

- Removal of the DC component present in the real Gabor filters (Section 4.2.1).

- Selection of CLS of interest, in order to reduce the amount of information unrelated to the presence of architectural distortion in the orientation field (Section 4.2.2).

- Filtering of the orientation field incorporating both the orientation field angle and the result of CLS selection (Section 4.2.3).

- Emphasis on the node pattern using a new term in the sum of squared errors of the phase portrait model (Equation 2.10), in order to facilitate the detection of spiculated architectural distortion (Section 4.2.4).

- Modifications in the post-processing and detection phase to improve the localization of sites of architectural distortion (Section 4.2.5).

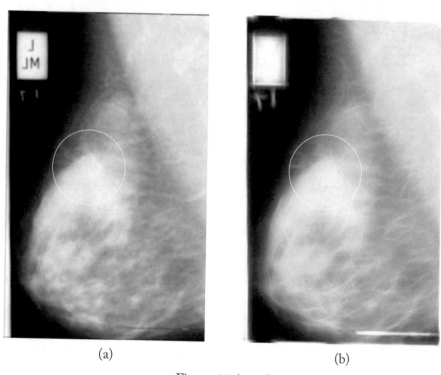

(a)   (b)

Figure 4.6 (cont.)

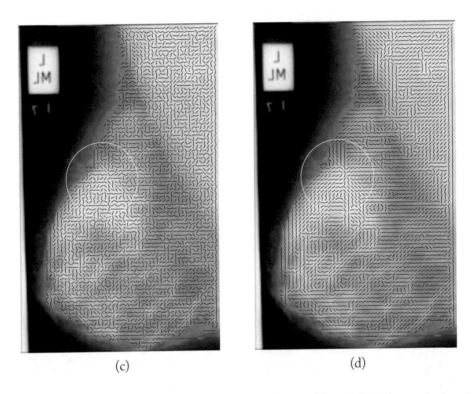

(c)   (d)

Figure 4.6: (a) Image 'mdb115' from the Mini-MIAS database. The circle indicates the location and extent of architectural distortion, as provided in the Mini-MIAS database. (b) Magnitude image after Gabor filtering. (c) Orientation field superimposed on the original image. Needles are drawn for every 5th pixel. (d) Filtered orientation field superimposed on the original image. Reproduced with permission from F. J. Ayres and R. M. Rangayyan. "Detection of architectural distortion in mammograms using phase portraits". In J. M. Fitzpatrick and M. Sonka, editors, Proceedings of SPIE Medical Imaging 2004: Image Processing, volume 5370, pages 587–597, San Diego, CA, February 2004. © SPIE

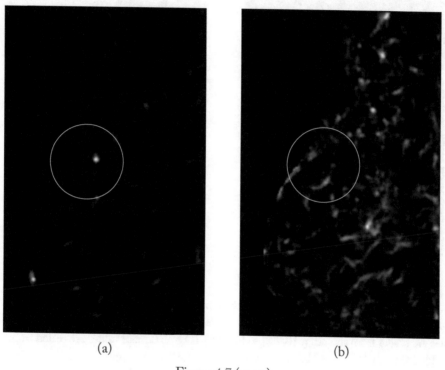

(a)   (b)

Figure 4.7 (cont.)

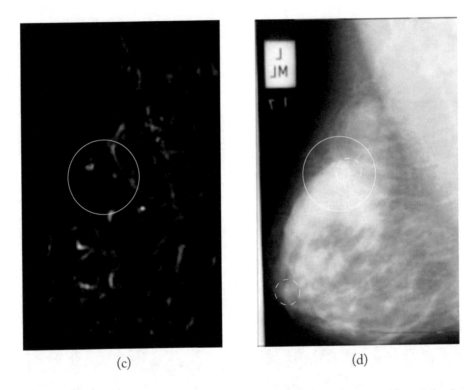

(c)   (d)

Figure 4.7: Phase portrait maps derived from the orientation field in Figure 4.6d, and the detection of architectural distortion. (a) Node map: values are scaled from the range [0, 84] to [0, 255]. (b) Saddle map: values are scaled from the range [0, 20] to [0, 255]. (c) Spiral map: values are scaled from the range [0, 47] to [0, 255]. (d) Detected sites of architectural distortion superimposed on the original image: the solid line indicates the location and spatial extent of architectural distortion as given by the Mini-MIAS database; the dashed lines indicate the detected sites of architectural distortion (one true positive and one false positive). Reproduced with permission from F. J. Ayres and R. M. Rangayyan. "Detection of architectural distortion in mammograms using phase portraits". In J. M. Fitzpatrick and M. Sonka, editors, Proceedings of SPIE Medical Imaging 2004: Image Processing, volume 5370, pages 587–597, San Diego, CA, February 2004. © SPIE

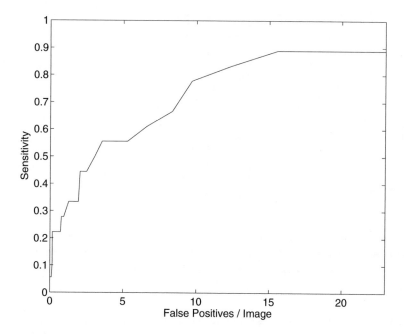

Figure 4.8: Free-response receiver operating characteristic (FROC) curve for the first method for the detection of architectural distortion. Reproduced with permission from F. J. Ayres and R. M. Rangayyan. "Detection of architectural distortion in mammograms using phase portraits". In J. M. Fitzpatrick and M. Sonka, editors, Proceedings of SPIE Medical Imaging 2004: Image Processing, volume 5370, pages 587–597, San Diego, CA, February 2004. © SPIE

The following sections provide details of the stages listed above. Section 4.2.6 presents the results obtained in the performance evaluation of this method.

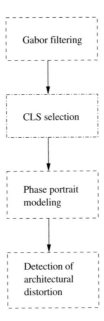

Figure 4.9: Flowchart for the second method for the detection of architectural distortion. Dashed line indicates procedures that are modified with respect to the first method, shown in Figure 4.5. Dash-and-dot line indicates a procedure that exists in the second method but not in the first method.

4.2.1 REMOVAL OF THE DC COMPONENT IN THE GABOR FILTERS

The Gabor filter has a nonzero magnitude response at the origin of the frequency plane (DC frequency). Consequently, the low-frequency components of the mammographic image being filtered may influence the result of the Gabor filter. Such influence will not affect the computation of the orientation field angle since the same influence will appear at all angles. However, the nonzero DC response will cause the orientation field magnitude to exhibit values that are affected by the low-frequency content of the image. It is desirable to reduce the influence of the low-frequency components of the mammographic image in the orientation field magnitude, since the low-frequency components are not related to the presence of oriented structures in the image. Therefore, the mammographic image is high-pass filtered prior to the extraction of the orientation field. The high-pass filtering step is achieved by computing the difference between the original mammographic image and a low-pass-filtered version of the same image. The low-pass filter employed is a Gaussian filter with $\sigma_{LPF} = \sigma_y$ (as defined in Section 1.2.2) and unit gain at the origin of the frequency plane.

4.2.2 SELECTION OF CURVILINEAR STRUCTURES

The Gabor filter bank is sensitive to linear structures, such as spicules and fibers. However, the filter bank also recognizes strong edges in the image as oriented features. Examples of strong edges in mammographic images are the edges of the pectoral muscle, edges of the parenchymal tissue, and vessel walls. Strong edges around the fibroglandular disk [59] may be of interest in the detection of focal retraction, a particular form of architectural distortion [60]. Nevertheless, in our method, it is desirable that only linear structures related to fibroglandular tissue are identified as oriented features.

The importance of identifying CLS for the detection of architectural distortion lies in the amount of information present in the configuration of the CLS. The analysis of CLS present in mammograms may improve the performance of algorithms for the detection of spiculated masses and architectural distortion, as suggested by Zwiggelaar *et al.* [16]. In a complementary manner, CLS could be identified and suppressed from mammographic images to facilitate the detection of masses [61].

The method for the selection of CLS implemented in this work includes three stages: segmentation of the breast area, detection of core CLS pixels, and rejection of CLS pixels at sites with a strong gradient [58].

The breast area is first segmented by the thresholding method of Otsu [62]. Although improved methods for the delineation of the breast boundary using active contour models have been proposed by Ferrari *et al.* [63], a simpler approach as above is adequate for the present application. Pixels outside the breast area are removed from further consideration at this stage.

The core CLS pixels are detected using the nonmaximal suppression (NMS) technique [64] applied to the magnitude image. The NMS algorithm identifies the core CLS pixels by comparing each pixel in the magnitude image with its neighbors along the direction that is perpendicular to the local orientation field angle; see Figure 4.10. If the pixel under investigation has a larger magnitude value than the corresponding neighbors, the pixel is considered to be a core CLS pixel. Nonmaximal suppression is a common step in many edge detectors (e.g., the Canny edge detector [65]), and Zwiggelaar *et al.* [16] used NMS, as described in this section, for the detection of CLS pixels.

The core CLS pixels associated with the presence of strong gradients are rejected. The rejection procedure implemented in this work follows the pixel rejection criteria proposed by Karssemeijer and te Brake [15], in the context of the detection of spiculated lesions. The gradient of the mammographic image is obtained using the first derivative of a Gaussian with a standard deviation of five pixels (1 mm). The direction of the gradient is computed, and compared to the direction of the orientation field, for each core CLS pixel. The core CLS pixel is discarded if the difference between the direction of the orientation field and the direction perpendicular to the gradient is less than $\pi/6$. It is assumed that the presence of a strong gradient causes a ripple in the magnitude image, leading to an erroneous detection of a CLS.

Curvilinear structures present within the fibroglandular disk will exhibit reduced contrast as compared to similar CLS outside the fibroglandular disk. As a consequence, CLS within the fibroglandular disk will present smaller magnitude field values than those of CLS outside the fi-

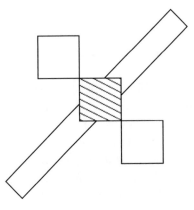

Figure 4.10: NMS technique: the elongated rectangle denotes the presence of a CLS, whereas the squares denote pixels along a direction perpendicular to the CLS orientation. The shaded square indicates a core CLS pixel.

broglandular disk. In order to assign the same weight to all CLS, independent of location, the magnitude field $M(x, y)$ is replaced by an image composed of the core CLS pixels, $M_{CLS}(x, y)$, defined as follows:

$$M_{CLS}(x, y) = \begin{cases} 1 & \text{if the pixel at } (x, y) \text{ is a core CLS pixel} \\ 0 & \text{otherwise.} \end{cases} \quad (4.4)$$

This procedure ensures that important CLS with low contrast, such as spicules within the fibroglandular disk, are not missed by the algorithm. Figures 4.11 and 4.12 show the results of CLS selection with a full mammogram and an ROI, respectively.

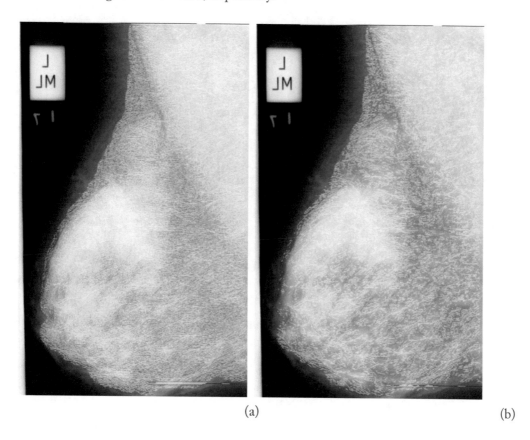

(a)   (b)

Figure 4.11: NMS and CLS results overlaid on a full mammographic image. (a) NMS results. (b) CLS results (after CLS rejection).

The image $M_{CLS}(x, y)$ conveys information on the presence of CLS. The magnitude of the detected CLS is of lesser importance, since the presence of architectural distortion is indicated by the geometrical arrangement of the CLS, rather than their intensity.

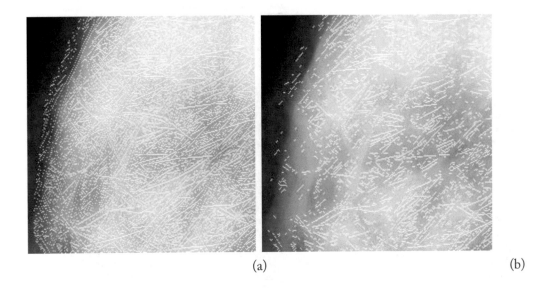

(a)   (b)

Figure 4.12: NMS and CLS results overlaid on a mammographic ROI. (a) NMS results. (b) CLS results (after CLS rejection).

4.2.3   FILTERING AND DOWNSAMPLING THE ORIENTATION FIELD

The orientation field is filtered and downsampled in a manner similar to that described in Section 4.1.1, but replacing the definition of $s(x, y)$ and $c(x, y)$ by the following equations:

$$s(x, y) = M_{CLS}(x, y) \sin[2\theta(x, y)] \tag{4.5}$$

and

$$c(x, y) = M_{CLS}(x, y) \cos[2\theta(x, y)]. \tag{4.6}$$

Also, the orientation field magnitude is obtained as

$$M(x, y) = (h * M_{CLS})(x, y), \tag{4.7}$$

with $h(x, y)$ as in Equation 4.1. The resulting angle and magnitude fields have a resolution of 0.8 mm/pixel.

4.2.4   ESTIMATING THE PHASE PORTRAIT MAPS

A penalty term is included to facilitate the detection of stellate (spiculated) architectural distortion. The penalty term is given by

$$\epsilon_{\text{node}}^2(\mathbf{A}) = 1000 \left[(a - d)^2 + 4bc \right]^2, \tag{4.8}$$

where

$$\mathbf{A} = \begin{bmatrix} a & b \\ c & d \end{bmatrix}, \tag{4.9}$$

which associates a higher penalty with configurations of the matrix $\mathbf{A}$ that deviate from a stellate node appearance. The sum of the squared error is given by

$$\epsilon^2(\mathbf{A}, \mathbf{b}) = \sum_x \sum_y M(x, y) \sin^2(\theta(x, y) - \phi(x, y | \mathbf{A}, \mathbf{b})) + \epsilon_{\text{node}}^2(\mathbf{A}), \tag{4.10}$$

where the range of the summation indices x and y is the set of the pixel locations within the analysis window. Estimates of $\mathbf{A}$ and $\mathbf{b}$ that minimize $\epsilon^2(\mathbf{A}, \mathbf{b})$ are obtained as follows:

1. Initial estimates $\mathbf{A}_{SA}$ and $\mathbf{b}_{SA}$ of $\mathbf{A}$ and $\mathbf{b}$ are obtained through the minimization of $\epsilon^2(\mathbf{A}, \mathbf{b})$ using simulated annealing [52].

2. A nonlinear least-squares algorithm [66] is used to refine the estimates $\mathbf{A}_{SA}$ and $\mathbf{b}_{SA}$ obtained in the previous step, producing the optimal estimates $\mathbf{A}_{opt}$ and $\mathbf{b}_{opt}$.

   The type of phase portrait is determined by the eigenvalues of $\mathbf{A}_{opt}$; the fixed-point location is given by Equation 2.5, using the optimal values $\mathbf{A}_{opt}$ and $\mathbf{b}_{opt}$. A vote is cast in the corresponding phase portrait map, at the location of the fixed point, if the following conditions are met:

- Both eigenvalues of **A** are not zero, in order to prevent numerical instabilities in the computation of $\mathbf{A}^{-1}$.

- The distance from the center of the analysis window to the fixed-point location is less than 20 pixels (16 mm).

This process is repeated for every position of the analysis window.

4.2.5   DETECTION OF ARCHITECTURAL DISTORTION

In this method, we use a procedure to detect and localize sites of architectural distortion, using only the node map, as follows:

1. A Gaussian filter with an empirically determined standard deviation of 6 pixels (4.8 mm) is applied to the node map in order to reduce noise. This step leads to real values in the node map, instead of integers equal to the the numbers of votes cast.

2. The filtered node map is processed with a morphological gray-scale opening procedure [57] with a circular structuring element of radius 10 pixels (8 mm) to eliminate peaks in the filtered node map that are closer than 8 mm to a locally dominant peak.

3. The peaks of the resulting image are detected, and a threshold is applied to eliminate false positives. The remaining peaks, if any, indicate the potential sites of architectural distortion.

The threshold value influences the sensitivity of the method and the number of false positives per image. A high threshold value reduces the number of false positives, but it also reduces the sensitivity. A low threshold value increases the number of false positives. Various values of the threshold may be used to generate an FROC curve [67]. A detected site of architectural distortion is considered to be a true-positive detection if its location is within the extent of architectural distortion given in the Mini-MIAS database [42].

4.2.6   RESULTS OF THE SECOND METHOD

The proposed methodology was applied to 19 mammograms exhibiting architectural distortion, from the Mini-MIAS database [42]. All images with architectural distortion in the MIAS database were used in this study. The mammograms are MLO views, digitized to $1,024 \times 1,024$ pixels at a resolution of 200 μm and 8 bits/pixel.

The application of the Gabor filtering method of Section 1.2.2 is illustrated in Figures 4.13b and 4.14a, which correspond to the magnitude and the angle components, respectively, of the orientation field of image 'mdb115' (Figure 4.13a). Observe that the directional structures present in the mammogram are enhanced in the magnitude image. Also, notice that the filtered orientation field angle (Figure 4.14b) is smoother and more coherent as compared to the original orientation field angle (Figure 4.14a). As a result, the pattern of architectural distortion is displayed better in

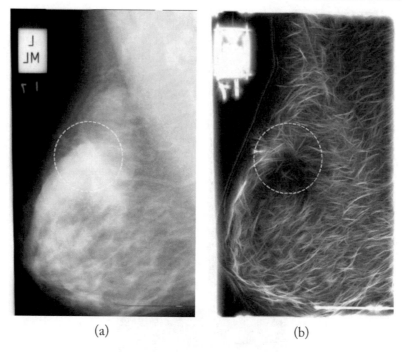

(a)   (b)

Figure 4.13: (a) Image 'mdb115' from the Mini-MIAS database. The dashed circle indicates the location and extent of architectural distortion, as provided in the database. (b) Magnitude image after Gabor filtering and removal of the DC component. Reproduced with permission from R. M. Rangayyan and F. J. Ayres. "Gabor filters and phase portraits for the detection of architectural distortion in mammograms". Medical and Biological Engineering and Computing, 44:883–894, August 2006. © Springer

the angle component of the orientation field: there is a convergence of lines towards the center of the indicated site of architectural distortion in Figure 4.14b.

The architectural distortion present in the mammogram 'mdb115' has a stellate or spiculated appearance. As a consequence, a large number of votes have been cast into the node map, at a location close to the center of the distortion, as seen in Figure 4.14c.

Small sites of architectural distortion may result in a proportionately lower number of votes cast, and hence reduce the detectability of the lesion. Figures 4.15 and 4.16 illustrate this phenomenon. It can be observed in Figure 4.15b and 4.16b that the pattern of architectural distortion is virtually indistinct from the remaining areas of the breast. As a consequence, the node map does not exhibit a prominent peak within the site of architectural distortion.

The FROC curve was derived by varying the threshold level in the detection step; the result is shown in Figure 4.17. A sensitivity of 84% was obtained at 7.8 false positives per image, improving upon the performance of the first method.

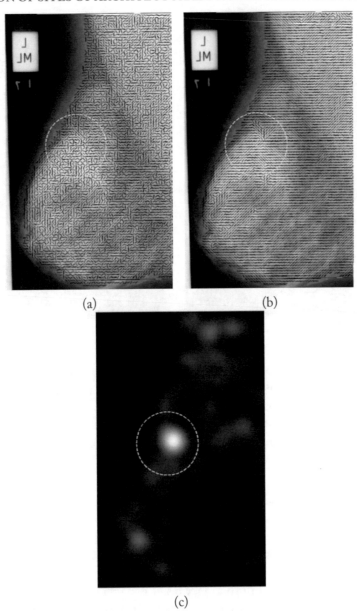

Figure 4.14: (a) Orientation field superimposed on the image 'mdb115' from the Mini-MIAS database. Needles are drawn for every 10th pixel. (b) Filtered orientation field superimposed on the original image. (c) Filtered node map: values are scaled from the actual range of [0, 0.53] to [0, 255]. The dashed circle indicates the site of architectural distortion. Reproduced with permission from R. M. Rangayyan and F. J. Ayres. "Gabor filters and phase portraits for the detection of architectural distortion in mammograms". Medical and Biological Engineering and Computing, 44:883–894, August 2006. © Springer

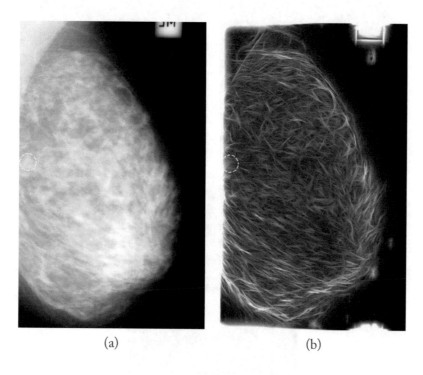

(a)   (b)

Figure 4.15: (a) Image 'mdb130' from the Mini-MIAS database. The dashed circle indicates the location and extent of architectural distortion, as provided in the database. (b) Magnitude image after Gabor filtering. Reproduced with permission from R. M. Rangayyan and F. J. Ayres. "Gabor filters and phase portraits for the detection of architectural distortion in mammograms". Medical and Biological Engineering and Computing, 44:883–894, August 2006. © Springer

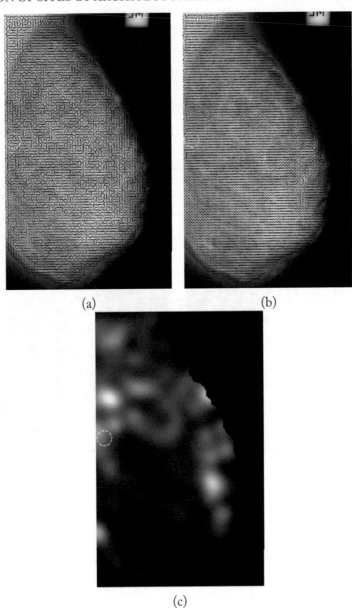

(a)   (b)

(c)

Figure 4.16: (a) Orientation field superimposed on the image 'mdb130' from the Mini-MIAS database. Needles are drawn for every 10th pixel. (b) Filtered orientation field superimposed on the original image. (c) Filtered node map: values are scaled from the actual range of [0, 0.24] to [0, 255]. The dashed circle indicates the site of architectural distortion. Reproduced with permission from R. M. Rangayyan and F. J. Ayres. "Gabor filters and phase portraits for the detection of architectural distortion in mammograms". Medical and Biological Engineering and Computing, 44:883–894, August 2006. © Springer

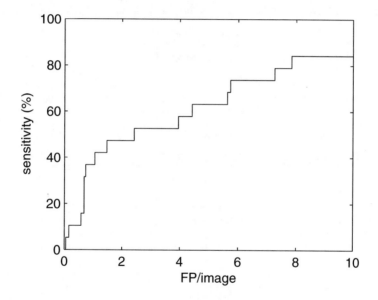

Figure 4.17: FROC curve for the second method for the detection of architectural distortion. FP = false positives. Reproduced with permission from R. M. Rangayyan and F. J. Ayres. "Gabor filters and phase portraits for the detection of architectural distortion in mammograms". Medical and Biological Engineering and Computing, 44:883–894, August 2006. © Springer

4.3   THE THIRD METHOD: SHAPE-CONSTRAINED PHASE PORTRAIT MODEL

The third and final variant of the method for the detection of architectural distortion [68, 69] (see the flowchart in Figure 4.18) incorporates the following modifications:

- The matrix $\mathbf{A}$ in Equation 2.7 is replaced by a symmetric matrix model (Section 4.3.1).

- The post-processing and detection phase incorporates the condition number of $\mathbf{A}$ into the vote-casting procedure (Section 4.3.2).

  The results of evaluation of performance of the third method are presented in Section 4.3.3.

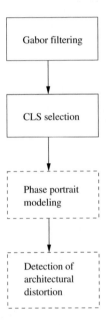

Figure 4.18: Flowchart for the third method for the detection of architectural distortion. Solid line indicates procedures that remain unmodified with respect to the second method, shown in Figure 4.9. Dashed line indicates procedures that are modified with respect to the second method method.

4.3.1   ADOPTION OF A SYMMETRIC MATRIX A

The general model in Equation 2.6 may lead to synthetic orientation fields that are unsuitable for the description of practical orientation fields in a specific application of interest. If the eigenvalues of $\mathbf{A}$ differ significantly in magnitude, or if the inner angle between the eigenvectors of $\mathbf{A}$ is small, the orientation field generated by the phase portrait model may degenerate into a pattern comprised of almost parallel lines.

The condition number of a matrix is the ratio between its largest and smallest singular values, and it is related to the degree of numerical precision achievable in the process of inverting the matrix. For a symmetric matrix, the singular values are equal to the absolute values of the eigenvalues. The condition number is always greater than or equal to one. A high condition number of a given matrix indicates that the corresponding inverse matrix will be highly sensitive to small perturbations in the original matrix. Thus, the geometrical focus of the pattern (given by the fixed-point location) cannot be established accurately. A singular matrix will have an infinite condition number, and it is related to a phase portrait pattern without a fixed point. (Therefore, in a practical application, a high condition number may indicate that the orientation field under analysis has no significant geometrical focus.) The examples in Figures 4.19 and 4.20 illustrate how the condition number of the model matrix $\mathbf{A}$ influences the appearance of the corresponding orientation field.

The condition number can, therefore, be incorporated into our algorithm for the detection of architectural distortion as a measure of the suitability of the related phase portrait pattern: Figure 4.21 illustrates examples to support this argument. In detail A, the line segments or needles (representing the orientation field within the observation window) convey the impression of convergence to a focal point, which, in this case, is within the site of architectural distortion. The condition number associated with the orientation field in detail A is 2.05, indicating that the eigenvalues of $\mathbf{A}$ have comparable magnitudes. The orientation field observed in detail B is comprised of almost a constant angle (parallel lines), and it does not give the impression of convergence. The corresponding value of the condition number is 55.44, denoting a significant imbalance between the magnitudes of the eigenvalues of $\mathbf{A}$.

In the present (third) method, the phase portrait model is modified to impose orthogonality of the eigenvectors. The matrix $\mathbf{A}$ in Equation 2.7 is replaced by the following symmetric matrix model:

$$\mathbf{A} = \begin{bmatrix} a & b \\ b & c \end{bmatrix}. \tag{4.11}$$

Then, the eigenvectors of $\mathbf{A}$ are mutually orthogonal, and the eigenvalues of $\mathbf{A}$ are real-valued [70]. Since the eigenvalues of the constrained matrix $\mathbf{A}$ are real-valued, only node and saddle phase portraits can be generated using this model. The exclusion of spiral patterns is appropriate for the present application to mammograms since such patterns were not observed in the test dataset.

4.3.2 VOTE CASTING AND DETECTION

In the vote-casting phase, unsuitable phase portrait models are rejected based on the condition number of $\mathbf{A}$: a phase portrait model is deemed to be unsuitable for further analysis if the condition number κ is greater than 3.0. The limit on the condition number ensures that the precision in the determination of the fixed-point location will not be affected by ill-conditioning of $\mathbf{A}$. Furthermore, as discussed in Section 4.3.1, a high condition number also indicates that no fixed-point location should be assigned to the orientation field pattern under analysis.

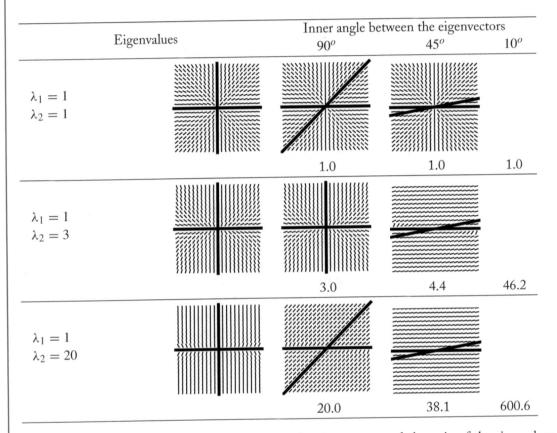

Figure 4.19: Influence of the inner angle between the eigenvectors and the ratio of the eigenvalues of the matrix $\mathbf{A}$ in the phase portrait model on the appearance of the orientation field. Only node patterns are displayed. The number shown under each orientation field is the condition number of the corresponding matrix $\mathbf{A}$ (see Section 4.3.1). Reproduced with permission from F. J. Ayres and R. M. Rangayyan. "Reduction of false positives in the detection of architectural distortion in mammograms by using a geometrically constrained phase portrait model". International Journal of Computer-Assisted Radiology and Surgery, 1(6):361–369, April 2007. © Springer

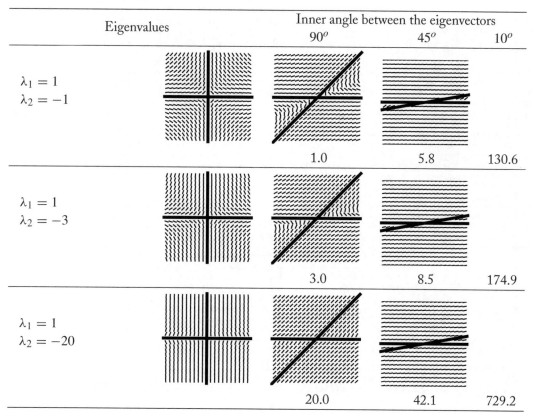

Figure 4.20: Influence of the inner angle between the eigenvectors and the ratio of the eigenvalues of the matrix **A** in the phase portrait model on the appearance of the orientation field. Only saddle patterns are displayed. The number shown under each orientation field is the condition number of the corresponding matrix **A** (see Section 4.3.1). Reproduced with permission from F. J. Ayres and R. M. Rangayyan. "Reduction of false positives in the detection of architectural distortion in mammograms by using a geometrically constrained phase portrait model". International Journal of Computer-Assisted Radiology and Surgery, 1(6):361–369, April 2007. © Springer

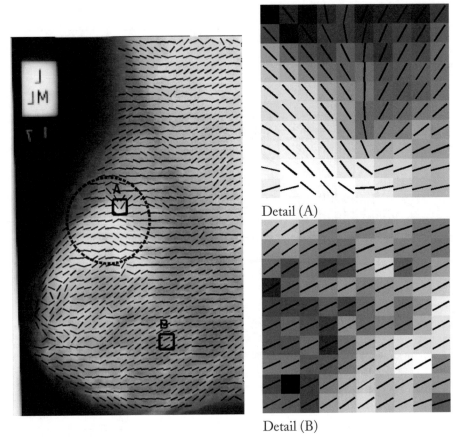

Detail (A)

Detail (B)

Figure 4.21: (left) Image 'mdb115' from the Mini-MIAS database with the orientation field superimposed (black lines or needles). The dashed circle indicates the site of architectural distortion. The image resolution is 800 μm/pixel. Needles are drawn every five pixels. Detail (A) from the orientation field (size 10×10 pixels): condition number = 2.05. Detail (B) from the orientation field (size 10×10 pixels): condition number = 55.44. In both details A and B, needles are drawn for every pixel. Reproduced with permission from F. J. Ayres and R. M. Rangayyan. "Reduction of false positives in the detection of architectural distortion in mammograms by using a geometrically constrained phase portrait model." International Journal of Computer-Assisted Radiology and Surgery, 1(6):361–369, April 2007. © Springer

The fixed point location is obtained for the estimated phase portrait model, at each position of the analysis window, and the distance from the center of the analysis window to the fixed-point location is computed. If the distance is less than five pixels (4 mm) or greater than 20 pixels (16 mm), the phase portrait under the current analysis window is rejected. This option for rejection is due to the following observations:

- Regions of architectural distortion are not expected to include a dense core or mass. However, the geometrical patterns associated with architectural distortion may be present at the periphery of the region and not extend to its center. Regardless, the central regions of sites of architectural distortion could contain patterns related to other areas of the breast superimposed in the mammogram; the lower limit of 4 mm in the distance-to-fixed-point constraint is included to prevent false positives due to such interfering patterns.

- The fixed-point location cannot be accurately determined when the distance from the center of the analysis window to the fixed-point location is large, with respect to the size of the analysis window.

If both conditions for the suitability of the phase portrait (condition number and distance to fixed point) are satisfied, then a vote is cast at the fixed-point location in one of the two phase portrait maps: node or saddle. The magnitude of the vote is given by the ratio of the measure of fit $\epsilon^2(\mathbf{A}, \mathbf{b})$ (defined in Equation 2.10) to the condition number of $\mathbf{A}$, thus emphasizing the isotropy of the phase portrait. Finally, the node map is filtered with a Gaussian window with $\sigma = 6$ pixels (4.8 mm) in order to group votes that are placed in close proximity (since votes related to the presence of a pattern of architectural distortion may be scattered over a small region, rather than grouping exactly at the same location in the corresponding phase portrait map), and analyzed to detect peaks that are related to the sites of architectural distortion. The saddle map was observed to lack discrimination across the mammograms tested and was eliminated from further consideration.

The results of application of the third method are illustrated in Figure 4.22, for a mammogram exhibiting architectural distortion (Figure 4.22a). It can be observed that the node map (Figure 4.22b) presents a distinct response at the site of architectural distortion.

4.3.3   RESULTS OF THE THIRD METHOD

Experiment 1:

The proposed method was applied to all of the 19 mammograms in the Mini-MIAS database [42] containing architectural distortion, with the inclusion of 41 normal mammograms from the same database. Figure 4.23 shows the resulting FROC curve (solid line). A sensitivity of 84% was obtained at 4.5 false positives per image, and a higher sensitivity of 95% was obtained at 9.9 false positives per image. A threshold value of $\kappa = 3$ on the condition number of $\mathbf{A}$ was used in this experiment, as described in Section 4.3.1.

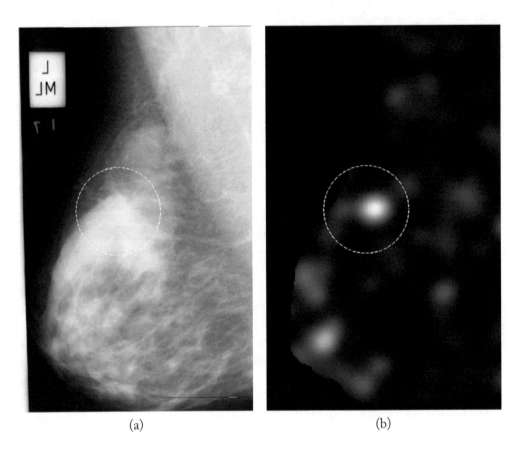

(a)   (b)

Figure 4.22: Results obtained using the third method. (a) Mammographic image ('mdb115' from the Mini-MIAS database) exhibiting architectural distortion. (b) Node map after Gaussian filtering ($\sigma = 4.8$ mm). The dashed circle indicates the site of architectural distortion. Reproduced with permission from F. J. Ayres and R. M. Rangayyan. "Reduction of false positives in the detection of architectural distortion in mammograms by using a geometrically constrained phase portrait model." International Journal of Computer-Assisted Radiology and Surgery, 1(6):361–369, April 2007. © Springer

The FROC curve obtained with the application of our second method for the detection of architectural distortion (Section 4.2), with no constraint on $\mathbf{A}$ related to symmetry or condition number, to the same dataset, is also shown in Figure 4.23 (dashed line). It can be observed that the detection performance of the third method is significantly better than that of the second method.

The effect of the threshold on the condition number on detection performance is illustrated in Figure 4.24. The FROC curves were obtained using different threshold values: $\kappa = 2$ (solid line) and $\kappa = 5$ (dashed line). The adoption of a lower threshold value than $\kappa = 3$ leads to a performance where the false-positive rate is reduced to eight per image at a sensitivity of 95%; however, at a lower sensitivity of 84%, the false-positive rate is 6.7 per image. A threshold of $\kappa = 5$ on the condition number results in a poorer performance than that obtained with either $\kappa = 2$ or $\kappa = 3$, as indicated by the FROC curve shown in Figure 4.24 (dashed line).

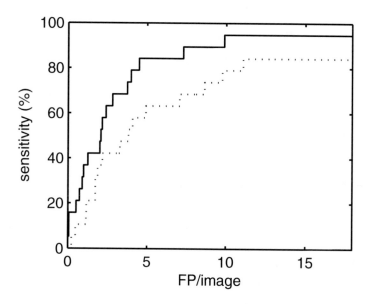

Figure 4.23: FROC curve for the detection of sites of architectural distortion, using the Mini-MIAS dataset with 19 cases of architectural distortion and 41 normal cases. FP = False positives. Solid line: third method. Dashed line: second method. The FROC curve for the third method was obtained using a threshold $\kappa = 3$ on the condition number of the matrix $\mathbf{A}$. Reproduced with permission from F. J. Ayres and R. M. Rangayyan. "Reduction of false positives in the detection of architectural distortion in mammograms by using a geometrically constrained phase portrait model." International Journal of Computer-Assisted Radiology and Surgery, 1(6):361–369, April 2007. © Springer

Experiment 2:

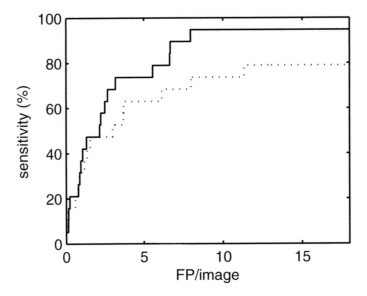

Figure 4.24: Effect of the threshold on the condition number on the detection performance of the third method. Solid line: $\kappa = 2$. Dashed line: $\kappa = 5$. Reproduced with permission from F. J. Ayres and R. M. Rangayyan. "Reduction of false positives in the detection of architectural distortion in mammograms by using a geometrically constrained phase portrait model." International Journal of Computer-Assisted Radiology and Surgery, 1(6):361–369, April 2007. © Springer

An independent evaluation of the performance of the proposed algorithm was conducted with iCAD — a company that manufactures CAD systems for mammography [71]. A dataset of 37 mammograms of biopsy-proven cases of architectural distortion was provided by iCAD, without any information regarding the presence and location of architectural distortion. In a correspondence after the analysis of the results, the following description of the dataset was provided: "The data set is comprised of 37 biopsy-proven architectural distortion images. A majority of the 37 cancers exhibit a spiculation signature. Approximately 50% of the lesions are obvious and should not be missed; an additional 25% are apparent, but not trivial. The remaining lesions are mainly architectural distortions without a definite spiculation signature — a very challenging class."

The third method was applied to the images provided. No changes were made to the method, and the same parameters as in Experiment 1 were used. The location and intensity of the top 20 peaks in the node map were recorded for each image. This information was sent to iCAD where it was compared with the ground-truth information. FROC analysis was conducted by iCAD, and the resulting FROC curve is given in Figure 4.25. A sensitivity of 81% was obtained at 10 false positives per image, and a higher sensitivity of 92% was achieved with 17 false positives per image. These results need to be analyzed in relation to the description of the dataset provided by iCAD: 25% of the cases do not have a definite signature of spiculation, and they would require a low detection threshold that incurs a high false-positive rate. It was observed by iCAD that the proposed method detected the locations of spiculated architectural distortion with good spatial accuracy, but it did not work well with images not exhibiting spiculated distortion. This indicates the need to incorporate other models in the procedure to address other patterns of architectural distortion.

4.4   REMARKS

In this chapter, three methods for the detection of architectural distortion were presented. Each method builds upon its predecessor: as a consequence, the statistical performance of the developed methods improves from the first method to the second, and from the second to the third method.

The adoption of a constrained phase portrait model with a symmetric matrix $\mathbf{A}$ and the incorporation of the condition number of $\mathbf{A}$ into the vote-casting procedure resulted in improved quality of information related to sites of architectural distortion in the node map, and a consequent reduction in the false-positive rate in the detection of architectural distortion. An experiment was conducted comparing the performance of the third shape-constrained method with the second method, where it was demonstrated that the shape-constrained method yields better performance. A blind experiment was also conducted with mammograms provided by iCAD: the results of this experiment indicate that the third method described in this chapter is promising in the detection of spiculated architectural distortion.

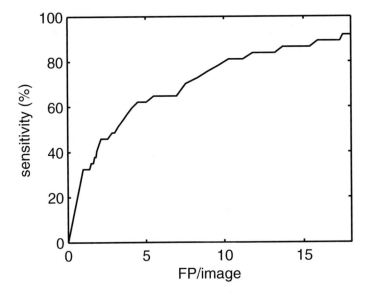

Figure 4.25: FROC curve for the detection of sites of architectural distortion (third method) with the dataset of 37 cases of architectural distortion provided by iCAD. Reproduced with permission from F. J. Ayres and R. M. Rangayyan. "Reduction of false positives in the detection of architectural distortion in mammograms by using a geometrically constrained phase portrait model." International Journal of Computer-Assisted Radiology and Surgery, 1(6):361–369, April 2007. © Springer

APPENDIX A

Computer-aided diagnosis of breast cancer

A.1 SCREENING FOR BREAST CANCER

Breast cancer is the most frequently diagnosed cancer in women. According to the National Cancer Institute of Canada, the lifetime probability of developing breast cancer is one in 8.9, and the lifetime probability of death due to breast cancer is one in 26.8 [72]. Breast cancer has the highest prevalence among all cancers in the female population, with 1.0% of all women living with the disease [72].

Early detection of breast cancer is of utmost importance: localized cancer leads to a five-year survival rate of 97.5%, whereas cancer that has spread to distant organs has a five-year survival rate of only 20.4% [73]. Breast self-examination is not adequate: many studies indicate that there is no evidence of a reduction in the mortality rate from breast cancer in women who practice regular breast self-examination, compared to those who do not [74, 75].

Mammography is, at present, the best available examination for the detection of early signs of breast cancer [74]. It can reveal pronounced evidence of abnormality, such as masses and calcifications, as well as subtle signs such as bilateral asymmetry and architectural distortion [76]. Mammographic screening has been shown to be effective in reducing breast cancer mortality rates: screening programs have reduced mortality rates by 30% to 70% [77], [78, chapter 19]. Cady and Chung [79] discuss the value of mammographic screening programs, highlighting the reduction in mortality achieved by several screening programs in Sweden, the Netherlands, the United Kingdom, Finland, Italy, and the United States. The drawbacks of screening are also discussed, such as the higher incidence of unnecessary biopsies, cost and quality of interpretation of mammograms versus the experience of the radiologists, and the psychological consequences of errors, such as the anxiety caused by a false-positive result and the wrongful reassurance provided by a false-negative test. It has been concluded that the benefits of screening surpass the drawbacks, and that the practice of mammographic screening must be encouraged and expanded.

However, interpreting screening mammograms is not easy: the sensitivity of screening mammography is affected by image quality and the radiologist's level of expertise. Another factor that affects a radiologist's performance is the high volume of cases examined in a screening program. The lack of expert radiologists to analyze mammograms in remote or rural areas is also a matter of concern. Bird *et al.* [80] estimated the sensitivity of screening mammography to be between 85% and 90%. Misinterpretation of breast cancer signs accounted for 52% of the errors, and overlooking signs corresponded to 43% of the missed abnormalities. In a study by van Dijck *et al.* [81], minimal signs of abnormalities were found to be present on screening mammograms taken previously in many cases of screen-detected cancers. Double reading of screening mammograms was found to provide greater sensitivity than single reading without increasing recall rates, in a comparative analysis by Blanks *et al.* [82], but the manpower required may render such an approach impractical.

A.2   MAMMOGRAPHIC SIGNS OF BREAST CANCER

A mammogram is an X-ray projection image of the breast. A summary description of the mammographic image acquisition process is as follows:

- The breast is compressed between two Plexiglas or plastic plates, and irradiated using low-energy X rays in a direction perpendicular to the compression plates (typical peak voltage: 24 – 26 kVp [83]).

- The X rays are partially absorbed and scattered by the breast tissue. The degree of absorption increases with the density of the material being irradiated.

- The intensity pattern of the X rays after traversing the breast is registered on a film. Recently, digital acquisition systems have been developed, where the X-ray intensity is measured by high-resolution image sensors and recorded digitally.

The projection nature of mammography results in an image where the anatomical features of the breast are superimposed. Compression of the breast during the imaging process assists in reducing the superposition of mammary structures by spreading them, and also aids in the selection of optimal radiation parameters.

The presence of breast cancer is manifested in mammographic images as various possible signs of abnormality; a list of some signs of breast cancer is given below.

- **Calcifications**: Calcifications are deposits of calcium in the breast tissue that may indicate the presence of breast cancer. Calcium, being denser than normal breast tissues, will cause calcifications to appear in mammograms as bright objects. The possible appearances of calcifications are described in the BI-RADS lexicon [84] as follows: "Benign calcifications are usually larger than calcifications associated with malignancy. They are much coarser, often round with smooth margins and are much more easily seen. Calcifications associated with malignancy are usually very small and often require the use of a magnifying glass to see them well."

- **Masses**: The presence of breast cancer causes a desmoplastic reaction in the breast tissue, which can be observed in mammographic images as a bright (dense) object (although generally not as bright as calcifications). The BI-RADS describes the shape of a mass as round, oval, lobular, or irregular. The margin of a mass can be described as circumscribed (well-defined or sharply defined), microlobulated, obscured (when segments of the margin are hidden by the superposition of adjacent normal tissue), indistinct, or spiculated (lines, or spicules, are observed to be radiating from the margins of the mass).

- **Bilateral asymmetry**: The occurrence of cancer may cause a modification in the overall appearance of the breast in a mammographic image. This modification is evidenced by differences in the distribution of density or the organization of fibroglandular tissue between the left and right breasts, even when more pronounced signs of cancer are not present (such as calcifications and masses).

- **Architectural distortion**: A localized sign of abnormality, manifested by the presence of a distortion in the normal architecture of the breast with no mass visible. See Chapter 4 for more details on architectural distortion.

Due to the superposition of anatomical features in the mammographic image, and the subtle nature of early signs of breast cancer, the analysis of mammograms is considered to be a difficult task, requiring that radiologists specialize in reading mammograms in order to increase the detection rates of the disease.

The development of new algorithms for CAD of breast cancer is an active research field [36, 85, 86, 171, 172], particularly in regard to the detection of subtle abnormalities in mammograms, and in spite of the success of a few commercial CAD systems [71, 174] in the improvement of the rates of detection of breast cancer. A substantial record of research exists in the literature regarding the detection and classification of masses and calcifications. These problems are generally considered to be well studied, and new developments must meet or exceed the high standards of performance set by the existing algorithms. Furthermore, commercial CAD systems have achieved a satisfactory degree of effectiveness in the detection of masses and calcifications. Nevertheless, certain areas of research in CAD of breast cancer still demand attention.

A relatively small number of researchers (as compared to the number of researchers who have conducted works related to masses and calcifications) have concentrated their attention on the problem of detecting architectural distortion in the absence of a central mass. Most of the published efforts are directed toward a more general category of abnormalities, such as spiculated lesions, which encompasses some of the possible appearances of architectural distortion. Other lines of research that require more attention include the analysis of bilateral asymmetry, curvilinear structures (CLS), and breast density as a predictor of the risk of breast cancer. In a larger context, areas of interest related to CAD of breast cancer include the development of systems for content-based retrieval of mammograms, indexed atlases, and data-mining systems. Full-field digital mammography systems could facilitate the routine application of the techniques mentioned above. The following sections provide brief reviews of the areas mentioned above [173].

A.3   ENHANCEMENT OF MAMMOGRAMS

The enhancement of mammographic images could improve the accuracy of detection of early signs of breast cancer. For reviews on image enhancement in mammography, see Rangayyan [36], Morrow *et al.* [87], and Rangayyan *et al.* [88].

Diagnostic features in mammograms, such as masses and calcifications, may be small and have low contrast with respect to the surrounding breast tissues. Such attributes could render the diagnostic features hard to detect. Contrast enhancement techniques can improve the ability of a radiologist to perceive subtle diagnostic features, leading to earlier, more accurate diagnosis of breast cancer. Contrast enhancement can improve the quality of an otherwise unsatisfactory mammogram, as stated by Ram [89], who further indicated that the application of contrast enhancement techniques in a clinical situation may reduce the radiation dose by about 50%.

Traditional image enhancement techniques [36, 184] have been applied to radiography for more than three decades. Chan *et al.* [90] investigated the application of unsharp masking for digital mammography. Receiver operating characteristics (ROC) studies were conducted, and it was shown that unsharp masking improved the detectability of calcifications in digital mammograms. However, the method increased noise and caused some artifacts in the images.

Classical image enhancement techniques are often global transformations, i.e., techniques that do not adapt to the local information content in an image. There is significant variability in the size and shape of diagnostic features in mammograms, and classical techniques often perform poorly in enhancing various sections of a given mammogram. Therefore, it is necessary to devise adaptive contrast enhancement algorithms for mammographic images, where the transformation is adapted to the local context of the given image. Laine *et al.* [91] presented a method for nonlinear contrast enhancement based on multi-resolution representation and the use of dyadic wavelets.

Gordon and Rangayyan [92] were the first to report on the use of adaptive-neighborhood image processing to enhance mammographic image contrast. Rangayyan and Nguyen [93] defined a tolerance-based method for growing foreground regions that could have arbitrary shapes rather than square shapes. Morrow *et al.* [87] further developed this approach with a new definition of background regions. The adaptive-neighborhood contrast enhancement (ANCE) algorithm works as follows: each pixel in the digitized mammographic image is taken as the seed pixel in a region growing procedure. The region growing procedure identifies the set of pixels that are similar and connected to the seed pixel (called the foreground region), as well as a three-pixel wide ribbon of pixels surrounding the foreground region (called the background region). The new value of the seed pixel in the contrast-enhanced image is determined by using the contrast value between the foreground and the background regions. Figure A.1 illustrates the result of the ANCE algorithm applied to a mammogram displaying a cluster of calcifications.

Dhawan *et al.* [94] investigated the benefits of various contrast transfer functions in an algorithm for contrast enhancement using square adaptive neighborhoods. The evaluated contrast transfer functions included $\ln(1 + 3C)$, $1 - \exp(-3C)$, $\sqrt{C}$, and $\tanh(3C)$, where C is the original contrast. They found that while a suitable contrast function was important to bring out the features of interest in mammograms, it was difficult to select such a function. Dhawan and Le Royer [95] proposed a tunable contrast enhancement function for improved enhancement of mammographic features.

It is important to distinguish between the effect of enhancement algorithms on the detection of the presence of features such as microcalcifications in an image, as against their effects on the diagnostic conclusion about a subject. Some image enhancement techniques may improve the visibility of diagnostic features, but distort their appearance and shape characteristics, possibly leading to misdiagnosis [96].

Rangayyan *et al.* [88, 97] investigated the performance of their ANCE algorithm in increasing the sensitivity of breast cancer diagnosis by using ROC analysis and McNemar's tests [36]. A set of 78 screen-film mammograms of 21 difficult cases (14 malignant and seven benign) and another set of 222 screen-film mammograms of 28 interval cancer patients and six benign control cases were digitized with a resolution of about $4096 \times 2048 \times 10$-bit pixels, and subsequently processed with the ANCE algorithm. The original films, as well as the corresponding unprocessed and processed digital images, were presented to six experienced radiologists for an ROC analysis of the difficult-case set, and to three radiologists for analysis of the interval-cancer set. It was observed that the radiologists' performance improved with ANCE processing, with respect to both film and digital image reading, in terms of the area under the ROC curve. It was also observed that the diagnostic sensitivity was improved by the ANCE algorithm. McNemar's tests of symmetry indicated that the diagnostic confidence for the interval-cancer cases was improved by the ANCE technique with a high level of statistical significance ($p = 0.0001$ to 0.005), with no significant effect on the diagnosis of the benign control cases (p-values of 0.08 to 0.1).

Sivaramakrishna *et al.* [98] compared the performance of several contrast enhancement algorithms in a preference study. The compared algorithms were: ANCE [87], adaptive unsharp masking [99], contrast-limited adaptive histogram equalization [100], and wavelet-based enhancement [101]. In a majority of the cases with microcalcifications, the ANCE algorithm provided the most-preferred images. In the case of images with masses, the unenhanced (original) images were preferred in most of the cases.

A.4 SEGMENTATION OF MAMMOGRAMS AND ANALYSIS OF BREAST DENSITY

It has been observed that increased breast density is generally associated with a higher risk of development of cancer [102]. Many researchers have investigated computer methods for the assessment of the risk of development of breast cancer via automated analysis of breast density. Byng *et al.* [103] computed the skewness of histograms of 24×24 pixel (3.12×3.12 mm) sections of mammograms. An average skewness was computed for each image by averaging over all the section-based skewness measures of the image. Mammograms of breasts with increased fibroglandular density were observed to have histograms skewed toward higher density, resulting in negative skewness. On the other hand, mammograms of fatty breasts tended to have positive skewness. The fractal dimension of the breast

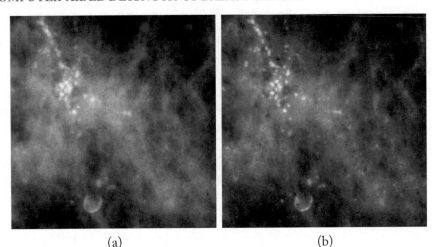

(a) (b)

Figure A.1: (a) Part of a mammogram with a cluster of calcifications; true size 43 × 43 mm. (b) Result of adaptive-neighborhood contrast enhancement. Reproduced with permission from W.M. Morrow, R.B. Paranjape, R.M. Rangayyan, and J.E.L. Desautels, "Region-based contrast enhancement of mammograms", IEEE Transactions on Medical Imaging, 11(3):392–406, 1992. © IEEE.

image was also computed: the image was interpreted as a relief map and the fractal dimension was computed using the box-counting method. The skewness and the fractal dimension measures were found to be useful in predicting the risk of development of breast cancer.

Caulkin *et al.* [104] observed that breast cancer occurs more frequently in the upper and outer quadrants of the breast, and that the majority of cancers are associated with glandular rather than fatty tissues. Therefore, the detection of different anatomical structures in the breast (such as fatty tissue, the fibroglandular disc, and the pectoral muscle), could facilitate the analysis of the risk of development of breast cancer, as well as the detection of early breast cancer. However, most of the proposed techniques for CAD of breast cancer analyze the whole mammogram, without considering the observation that signs of breast cancer may have different appearances in different regions. Based upon these observations, some researchers have proposed methods to segment and also to model mammograms in terms of anatomical regions.

Karssemeijer [105] used the Hough transform to identify the pectoral muscle as a straight-line edge in the mammogram. Ferrari *et al.* [106] proposed two methods for the identification of the pectoral muscle in mammograms. The first method is a variant of Karssemeijer's method, which employs the Hough transform and filtering applied to the accumulator cells. However, the hypothesis of a straight line for the representation of the pectoral muscle edge is not always valid, and may impose limitations on subsequent stages of image analysis. The second method proposed by Ferrari *et al.* [106], based upon directional filtering using Gabor wavelets, overcomes this limitation.

Saha *et al.* [107] employed scale-based fuzzy connectivity methods to segment dense regions from fatty regions in mammograms. The segmented dense and fatty regions were quantified by measuring the respective area and total density, and a set of features was derived from these measures. The features were linearly correlated between the MLO and the cranio-caudal (CC) views in order to demonstrate the inter-view similarity. The precision in the segmentation was measured by comparing the automatically segmented contours of the dense regions with manually delineated references drawn by experienced radiologists. The method was tested on 60 cases, each case including the MLO and CC projections. The method was found to be robust in the segmentation of dense regions, and the authors observed the density features to be strongly correlated between the MLO and CC views.

Several authors have reported on techniques for the delineation of the breast boundary. Ferrari *et al.* [63] developed a method for the identification of the breast boundary using active contour models, in which the mammogram is first contrast-enhanced and thresholded, producing an initial chain-code representation of the breast boundary. The final boundary is obtained by the application of a specially tailored active contour model algorithm. The method was applied to 84 MLO mammograms from the Mini-MIAS database [42]. The evaluation of the breast contours obtained by this method was performed based upon the percentage of false-positive and false-negative pixels, in comparison to contours that were manually drawn by a radiologist. The average false-positive and false-negative rates were 0.41% and 0.58%, respectively.

Ferrari *et al.* [59] proposed a method to segment the fibroglandular disc in mammograms based upon the Gaussian mixture model. In this method, prior to the detection of the fibroglandular disc, the breast boundary and the pectoral muscle are detected using other methods developed by the authors [63, 106] (mentioned above). The fibroglandular disc is detected by defining a breast density model. The parameters of the model are estimated using the expectation-maximization algorithm and the minimum-description-length principle. A qualitative assessment of the segmentation results, performed by an experienced radiologist, resulted in 64.3% of the results being rated as excellent, 16.7% rated as good, 10.7% rated as average, 4.7% rated as poor, and only 3.6% of the results as failed segmentation.

The prior segmentation and removal of the pectoral muscle from mammograms (MLO views) as well as the detection of the breast boundary and removal of artifacts outside the breast region in mammograms (as described above) should lead to improved analysis of the density of the breast. Furthermore, accurate delineation of the fibroglandular disc and statistical representation of the various types of tissue within the breast using a Gaussian mixture model [59] should improve the accuracy and extend the scope of analysis of breast density. The approach described above could lead to improvements in the prediction of the risk of development of breast cancer based upon screening mammograms.

A.5 DETECTION AND CLASSIFICATION OF MICROCALCIFICATIONS

The detection and classification of microcalcifications has been extensively studied, with many authors reporting on several successful approaches to this task. A recent survey by Cheng *et al.* [108] lists almost 200 references on computer-aided detection and classification of microcalcifications, including methods for the visual enhancement of microcalcifications, segmentation, detection, analysis of malignancy, and strategies for the evaluation of detection algorithms.

Shen *et al.* [109] proposed a method for the detection and classification of mammographic calcifications. The method starts with a multi-tolerance region-growing procedure for the detection of potential calcification regions and the extraction of contours. Shape features based on central moments, Fourier descriptors, and compactness are then extracted. Finally, a neural network is used for the classification of the feature vectors in order to distinguish between malignant and benign calcifications. The correct classification rates for benign and malignant calcifications were 94% and 87%, respectively [109]. In a related work [110] on the investigation of shape features and a more extensive analysis of classification of calcifications, a classification accuracy of 100% was obtained for both benign and malignant calcifications with a database containing 143 biopsy-proven calcifications (79 malignant and 64 benign).

Bankman *et al.* [111] reported on the use of a region-growing-based algorithm for the segmentation of calcifications that did not require threshold or window selection, and compared their algorithm to the aforementioned multi-tolerance method of Shen *et al.* as well as to an active contours method. A theoretical analysis of the computational complexity of each method was presented, along with computer execution times for comparison. The authors found that all of the three methods had similar statistical performance; however, their own algorithm outperformed the other methods in terms of computational effort.

Strickland [112] developed a two-stage method based on wavelet transforms for the detection and segmentation of microcalcifications. In this method, the detection of calcifications is performed in the wavelet domain. The detected sites are enhanced in the wavelet domain, prior to the computation of the inverse wavelet transform. The appearance of microcalcifications is enhanced by this procedure; a threshold procedure suffices to segment the calcifications. The test database consisted of 40 mammograms, and a sensitivity of 91% at three false positives per image was obtained.

El-Naqa *et al.* [113] used support vector machines to detect microcalcification clusters. The algorithm was tested using 76 mammograms, containing 1, 120 microcalcifications. A sensitivity of 94% was reported, at one false positive per image. An improvement of the method was published by Wei *et al.* [114] using a relevance vector machine. A database of 141 mammograms containing microcalcifications was used to test the algorithm. The method achieved a sensitivity of 90% at one false positive per image. The statistical performance of the method was similar to that of the method of El-Naqa *et al.* [113], but the authors reported a 35-fold improvement in computational speed.

Yu *et al.* [115] used a wavelet filter for the detection of microcalcifications, and a Markov random field model to obtain textural features from the neighborhood of every detected calcification. The Markov-random-field-based textural features, along with three auxiliary textural features (the mean pixel value, the gray-level variance, and a measure of edge density), were used to reject false positives. The method was evaluated using 20 mammograms containing 25 areas of clustered microcalcifications. A sensitivity of 92% was obtained, at 0.75 false positive per image.

Yu and Guan [116] developed a technique for the detection of clustered microcalcifications that is comprised of two parts: detection of potential microcalcification pixels, and delineation of individual microcalcifications by the elimination of false positives. The first part involves the extraction of features based on wavelet decomposition and gray-level statistics, followed by a neural-network classifier. The detection of individual objects required a vector of 31 features related to gray-level statistics and shape factors, followed by a second neural-network classifier. A database of 40 mammograms containing 105 clusters of calcifications was used to assess the performance of the proposed algorithm: a sensitivity of 90% was attained with 0.5 false positive per image.

Soltanian-Zadeh *et al.* [117] compared four groups of features according to their discriminant power in separating microcalcifications into the benign and malignant categories. The microcalcifications were segmented using an automated method, and several features were extracted. Each feature belonged to one of the following four categories: multi-wavelet-based features, wavelet-based features, Haralick's texture features [118], and shape features. Within each group, a feature-selection procedure based on genetic algorithms was employed to identify the most-suitable features for use with a k-nearest-neighbor classification scheme. The classification performance of each group of features was then determined using ROC analysis. The area under the ROC curve obtained ranged from 0.84 to 0.89, and it was observed that the multi-wavelet features yielded the best performance, followed by the shape features.

Serrano *et al.* [119, 185] and Acha *et al.* [120] proposed a method for the detection of calcifications based upon the error of a two-dimensional (2D) adaptive linear prediction algorithm [121] applied to the mammographic image. The method is based upon the observation that a microcalcification can be seen as a point of nonstationarity in an approximately homogeneous region or neighborhood in a mammogram; such a pixel cannot be predicted well by the linear predictor, and hence leads to a high error. The algorithm detects and localizes calcifications, and a multi-tolerance region-growing algorithm [109] is employed to delineate each calcification. The results of this procedure are illustrated in Figure A.2 for a part of a mammogram with calcifications. It can be observed that the calcifications are correctly delineated in Figure A.2c, despite the poor contrast between the calcifications and the dense breast tissue in the background.

A.6   DETECTION AND CLASSIFICATION OF MASSES

Several techniques have been developed for the detection and classification of breast masses in mammograms. The commercial CAD systems available at present incorporate some of these techniques, and there is mounting evidence in the scientific literature that such systems perform adequately.

Brzakovic *et al.* [122] reported on the use of a fuzzy pyramid linking technique for mass localization and shape analysis for false-positive elimination. Evaluation of the method was carried out on a small database of 25 mammographic images, leading to a classification accuracy of 85%.

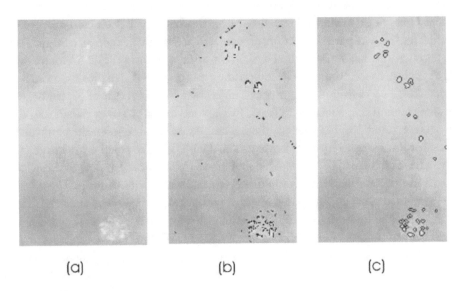

(a)   (b)   (c)

Figure A.2: (a) Mammogram section with malignant calcifications; 234 × 137 pixels with a resolution of 160 μm/pixel. (b) Seed pixels detected by thresholding the prediction error (marked in black). (c) Contours of the calcification regions detected by region growing from the seed pixels in (b). Reproduced with permission from C. Serrano, J.D. Trujillo, B. Acha, and R.M. Rangayyan, "Use of 2D linear prediction error to detect microcalcifications in mammograms", *CDROM Proceedings of the II Latin American Congress on Biomedical Engineering*, Havana, Cuba, 23–25 May 2001. © Cuban Society of Bioengineering.

Kegelmeyer *et al.* [123] proposed an algorithm for the detection of spiculated lesions that employed four Laws texture measures [124] and a new feature sensitive to stellate patterns. The test database consisted of 85 cases, with 49 normal cases and 36 positive cases; a total of 330 mammograms were used, with 68 lesions. A sensitivity of 97% was achieved at 0.28 false positive per image.

Karssemeijer and te Brake [15] developed a method for the detection of stellate patterns in mammograms, based on a statistical analysis of a map of the texture orientation in the mammographic images. The method for texture orientation analysis employs a multi-scale technique, and the orientation map is analyzed through the use of operators sensitive to stellate patterns. A sensitivity of 90% with one false positive per image was obtained in the detection of malignant stellate lesions and architectural distortion, using 31 normal cases and 19 cases with stellate lesions from the MIAS database [42]. In a related work, te Brake and Karssemeijer [125] presented an algorithm for the identification of masses that is an extension of their previous work on the detection of stellate patterns. The mass-detection algorithm identifies patterns of radial gradient vectors, rather than radial spiculations. A sensitivity of 75% was attained with one false positive per image, with a test database of 71 cases (132 mammograms) containing malignant tumors.

Sahiner *et al.* [127, 128] defined a "rubber-band straightening transform" (RBST) to map ribbons around breast masses in mammograms into rectangular arrays, and then computed Haralick's measures of texture [118]. The boundaries of 249 mammographic masses were automatically extracted. Haralick's texture measures individually provided classification accuracies of up to only 0.66, whereas the Fourier-descriptor-based shape factor defined by Shen *et al.* [110] gave an accuracy of 0.82 (the highest among 13 shape features, 13 texture features, and five run-length statistics). Each texture feature was computed using the RBST method [128] in four directions and for 10 distances. The full set of the shape factors provided an average accuracy of 0.85, the texture feature set provided the same accuracy, and the combination of shape and texture feature sets provided an improved accuracy of 0.89. These results indicate the importance of including features from a variety of perspectives and image characteristics in pattern classification.

Mudigonda *et al.* [130] proposed a method for the detection of masses in mammographic images based on the analysis of iso-intensity contour groups, and subsequent inspection of texture flow-field information to eliminate false positives. The test dataset consisted of 56 images from the Mini-MIAS database [42] including 30 benign lesions, 13 malignant cases, and 13 normals. The authors reported a sensitivity of 81% at 2.2 false positives per image. Figure A.4 illustrates the results of the procedure applied to an ROI of a mammogram containing two circumscribed benign masses. The method was also applied to the detection of masses in full mammographic images: the results are illustrated in Figure A.5 for a mammogram with a small malignant tumor.

Mudigonda *et al.* [129, 130] computed Haralick's texture measures using adaptive ribbons of pixels extracted around mammographic masses (see Figure A.3c), and used the features to distinguish malignant tumors from benign masses using linear discriminant analysis. The method was tested on a database of 39 mammographic images, including 16 circumscribed benign, four circumscribed malignant, 12 spiculated benign, and seven spiculated malignant masses. The authors reported a classification accuracy of 74.4% with an area under the ROC curve of $A_z = 0.67$. It was observed that restricting feature extraction to ribbons around the contours of masses improved the classification accuracy as compared to extracting the features over the entire regions of the masses.

Rangayyan *et al.* [131] proposed the use of shape factors and edge acutance (see Figure A.3d) for the classification of manually segmented masses as benign or malignant, and spiculated or circumscribed. An overall classification accuracy of 95% was obtained with a database of 54 mammographic images, including 16 circumscribed benign, seven circumscribed malignant, 12 spiculated benign, and 19 spiculated malignant masses.

Rangayyan *et al.* [126] introduced two new shape factors, spiculation index and fractional concavity (see Figure A.3b), and applied them for the classification of manually segmented mammographic masses. The combined use of the spiculation index, fractional concavity, and compactness yielded a benign-versus-malignant classification accuracy of 81.5%.

Li *et al.* [132] proposed a method for mass detection that employs a directional wavelet transform for multi-scale representation of the mammographic image, followed by segmentation of the mass at different scales, and the elimination of false-positive segments using shape analysis. A sensitivity of 91% with 3.2 false positives per image was obtained during the training phase of the proposed algorithm. The trained algorithm identified six of 10 subtle masses in a subsequent testing phase.

Zheng and Chan [133] devised an algorithm for the detection of masses that combines localized fractal analysis for preselection of suspicious regions, a multi-resolution Markov random field segmentation algorithm, and shape-based classification of segmented regions for reducing the number of false positives. The algorithm was evaluated using all of the 322 images in the Mini-MIAS database [42], and a sensitivity of 97.3% with 3.9 false positives per image was reported.

Liu *et al.* [134] formulated a multi-resolution procedure for the detection of spiculated lesions in digital mammograms. A multi-resolution representation of the mammographic image was obtained using a linear-phase, nonseparable, 2D wavelet transform. Pixel-based features were extracted at each resolution level, and the resulting feature maps were analyzed, from the coarsest to the finest resolution, to determine the sites of spiculated lesions. The algorithm was tested using a database of 19 spiculated lesions and 19 normal mammograms from the MIAS database, and a sensitivity of 100% with 2.2 false positives per image was reported.

Zwiggelaar *et al.* [135] introduced a technique to detect abnormal patterns of linear structures by detecting the radiating pattern of linear structures and/or the central mass expected to occur with spiculated lesions. Principal component analysis (PCA) was applied to a training set of mammograms including normal tissue patterns and spiculated lesions. The results of PCA were used to construct a basis set of oriented texture patterns, which was used to analyze radiating structures. A sensitivity of 80% was obtained at 0.23 false positive per image.

Guliato *et al.* [136] devised two methods for the segmentation of masses using fuzzy sets. The first method determines the boundary of a mass by region growing, after a fuzzy-set-based preprocessing enhancement step; the method yielded tumor boundaries that were consistent with manually drawn contours. The second method incorporated the fuzzy-set theory into the region-growing procedure, producing a fuzzy segmentation of the masses. The authors observed that the degree of inhomogeneity around the mass boundary correlated with the benign/malignant nature of the tumor, because malignant tumors are expected to have ill-defined margins. Using a measure of inhomogeneity of the result of fuzzy segmentation in the boundary region, the authors obtained a benign/malignant classification sensitivity of 80% with a specificity of 90%. Guliato *et al.* [137] also developed a method for combining multiple segmentation results into a single result, using fuzzy fusion operators. The resulting segmented regions were observed to be more consistent with the corresponding regions segmented by a radiologist than the individual results of segmentation.

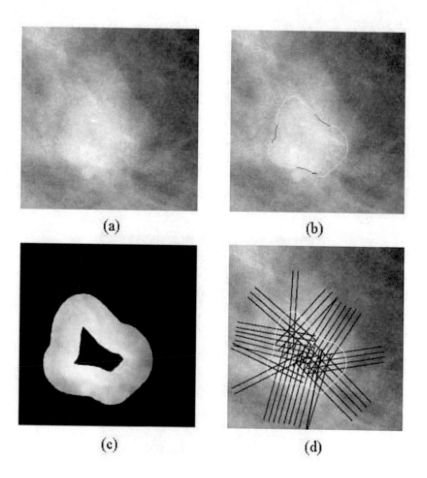

(a)   (b)

(c)   (d)

Figure A.3: (a) ROI of a benign mass. (b) ROI overlaid with the contour, demonstrating concave parts in black and convex parts in white. (c) Ribbon of pixels for the purpose of computing texture measures, derived by dilating the contour in (b). (d) Normals to the contour, shown at every tenth point on the contour, used for the computation of edge-sharpness measures. Reproduced with permission from H. Alto, R.M. Rangayyan, and J.E.L. Desautels, "Content-based retrieval and analysis of mammographic masses", *Journal of Electronic Imaging*, Vol. 14, No. 2, Article 023016, pp 1 – 17, 2005. © SPIE and IS&T.

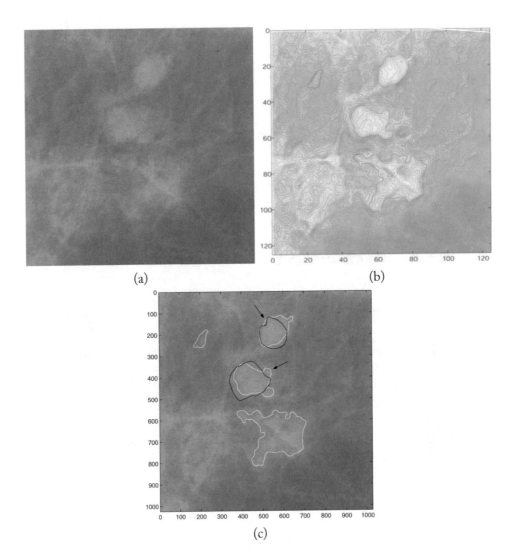

(a)   (b)

(c)

Figure A.4: (a) A $1,024 \times 1,024$-pixel section of a mammogram containing two circumscribed benign masses. Pixel size = 50 μm. Image width = 51 mm. (b) Groups of iso-intensity contours in the third multi-resolution version of the image in (a). (c) The contours (white) of two masses (indicated by arrows) and two false positives detected, with the corresponding contours (black) of the masses drawn independently by a radiologist. Reproduced with permission from N.R. Mudigonda, R.M. Rangayyan, and J.E.L. Desautels, "Segmentation and classification of mammographic masses", *Proceedings of SPIE Volume 3979, Medical Imaging 2000: Image Processing,* pp 55 – 67, 2000. © SPIE.

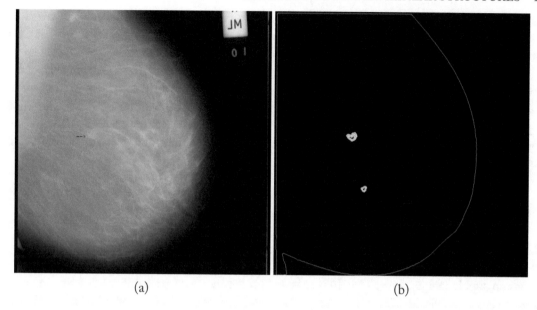

(a)   (b)

Figure A.5: (a) A mammogram (size 1, 024 × 1, 024 pixels, 200 μm per pixel) with a spiculated malignant tumor (pointed by the arrow, radius = 0.54 cm). Case mdb144 from the MIAS database. (b) Adaptive ribbons of pixels (white) and boundaries (black) of the regions retained in the mammogram after the false-positive analysis stage. The larger region corresponds to the malignant tumor; the other region is a false positive. Reproduced with permission from N.R. Mudigonda, R.M. Rangayyan, and J.E.L. Desautels, "Detection of breast masses in mammograms by density slicing and texture flow-field analysis", *IEEE Transactions on Medical Imaging,* 20(12): 1215 – 1227, 2001. © IEEE.

A.7   ANALYSIS OF CURVILINEAR STRUCTURES

The presence of CLS is an important factor in the detection of abnormalities in mammograms. The breast contains many structures that correspond mammographically to CLS, such as milk ducts, blood vessels, ligaments, parenchymal tissue, and edges of the pectoral muscle. Some lesions are characterized by the presence of certain types of CLS, such as spicules, in the mammographic image (for example, spiculated masses and architectural distortion), or by the asymmetric disposition of the oriented texture in the breast image. Conversely, some lesions, such as circumscribed masses, may be obscured by superimposed CLS; the resulting altered appearance could lead to misdiagnosis. Therefore, the ability to detect and classify CLS could enhance the performance of CAD algorithms.

Evans *et al.* [138] developed a method for statistical characterization of normal CLS in mammograms. Curvilinear structures were detected automatically, and six shape features were computed from each CLS. Principal component analysis was performed, and the two major dimensions were modeled using a Gaussian mixture model.

Wai *et al.* [139] proposed a method for the segmentation of CLS based on physical modeling of CLS in the breast. Qualitative experiments were conducted, and the authors reported that their method produced localized responses that are robust to the presence of noise.

Zwiggelaar *et al.* [16] investigated the performance of different methods for the detection and classification of CLS in mammograms, including: a line operator [22], orientated bins [37], steerable filters [26], and ridge detectors [38]. It was observed that the best method for CLS detection (line operator) yielded an area under the ROC curve of $A_z = 0.94$. Cross-sectional analysis of the detected profiles was performed, using PCA for dimensionality reduction, resulting in good discrimination between spicules and ducts ($A_z = 0.75$).

A.8   ANALYSIS OF BILATERAL ASYMMETRY

One of the cues used by radiologists to detect the presence of breast cancer is bilateral asymmetry, where the left and right breasts differ from each other in overall appearance in the corresponding mammographic images. Scutt *et al.* [43] reported on the increased probability of development of breast cancer associated with this finding. Miller and Astley [140] proposed a technique for the detection of bilateral asymmetry that comprised a semi-automated texture-based procedure for the segmentation of the glandular tissue, and measures of shape and registration cost between views for the detection of the occurrence of asymmetry. An accuracy of 86.7% was reported, on a test dataset of 30 screening mammogram pairs. In another report, Miller and Astley [141] presented a method for the detection of bilateral asymmetry based on measures of shape, topology, and distribution of brightness in the fibroglandular disk. The method was tested on 104 mammogram pairs, and a classification accuracy of 74% was obtained. Lau and Bischof [142] devised a method for the detection of breast tumors, using a localized definition of asymmetry that encompassed measures of brightness, roughness, and directionality. The method was evaluated using 10 pairs of mammograms where asymmetry was a significant factor in the radiologist's diagnosis. A sensitivity of 92% was obtained with 4.9 false positives per mammogram.

Ferrari *et al.* [30] developed a method for the analysis of asymmetry in mammograms using directional filtering with Gabor wavelets. In their method, the fibroglandular disk is segmented (Figure A.6 illustrates this step of the method), and the resulting image is decomposed using a bank of Gabor filters at different orientations and scales. The Karhunen-Loève transform is employed to select the principal components of the filter responses. Rose diagrams are computed from the phase images, and subsequently analyzed to detect the presence of asymmetry as characterized by variations in oriented textural patterns (see Figure A.7). A database of 80 images from the Mini-MIAS database containing 20 normal cases, 14 asymmetric cases, and six architectural distortion cases was used to evaluate the algorithm. The authors reported classification accuracy rates of up to 74.4%. The Gabor-filter-based method gives quantitative measures of the differences in the directional distribution of the fibroglandular tissue (pattern asymmetry). Rangayyan *et al.* [143] extended the method of Ferrari *et al.* [30] by including morphological measures quantifying differences in fibroglandular-tissue-covered area in the left and right breasts, which relate to size and shape. A sensitivity of 82.6% and a specificity of 86.4% were obtained in the detection of bilateral asymmetry.

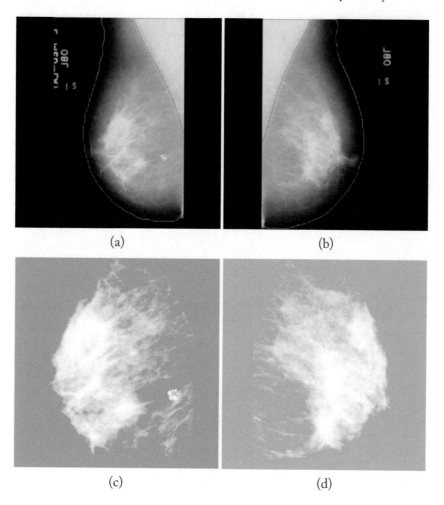

(a)   (b)

(c)   (d)

Figure A.6: Images mdb119 and mdb120 of a case of architectural distortion. (a) and (b) Original images ($1,024 \times 1,024$ pixels at $200~\mu$m/pixel). The breast boundary (white) and pectoral muscle edge (black) detected are shown. (c) and (d) Fibroglandular discs segmented and enlarged (512×512 pixels). Histogram equalization was applied to enhance the global contrast of each ROI for display purposes only. Reproduced with permission from R.J. Ferrari, R.M. Rangayyan, J.E.L. Desautels, and A.F. Frère, "Analysis of asymmetry in mammograms via directional filtering with Gabor wavelets", IEEE Transactions on Medical Imaging, 20(9): 953 – 964, 2001. © IEEE.

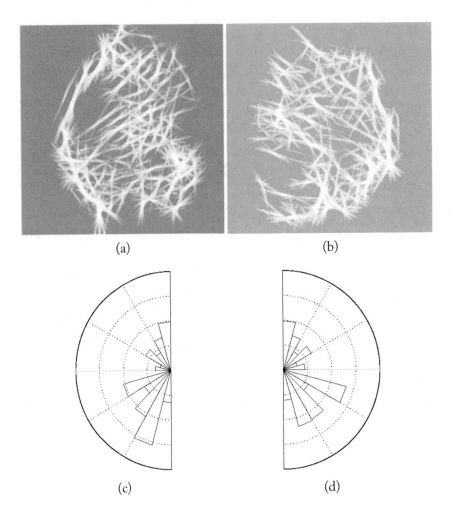

Figure A.7: Results of analysis of bilateral asymmetry for the case of architectural distortion in Figure A.6. (a) and (b) Magnitude images resulting from Gabor filtering and PCA. (c) and (d) Rose diagrams. The magnitude images were histogram-equalized for improved visualization. The rose diagrams have been configured to match the mammograms in orientation. Reproduced with permission from R.J. Ferrari, R.M. Rangayyan, J.E.L. Desautels, and A.F. Frère, "Analysis of asymmetry in mammograms via directional filtering with Gabor wavelets", *IEEE Transactions on Medical Imaging*, 20(9): 953 – 964, 2001. © IEEE.

A.9 DETECTION OF ARCHITECTURAL DISTORTION

Architectural distortion is one of the most commonly missed abnormalities in screening mammography. Improvements in the detection rate of architectural distortion could increase the rate of detection of early breast cancer, and reduce the morbidity and mortality due to breast cancer. We surmise that methods designed exclusively for the detection of architectural distortion can achieve better performance than the application of methods for the detection of spiculated masses, which may rely on the presence of a central mass.

Sampat *et al.* [144] employed filtering in the Radon-transform domain to enhance mammograms, followed by the use of radial spiculation filters to detect spiculated lesions. The algorithm was tested on 45 cases exhibiting spiculated masses, and 45 cases with the presence of architectural distortion. A sensitivity of 80% was obtained at 14 false positives per image in the detection of architectural distortion, and 91% at 12 false positives per image in the detection of spiculated masses.

The use of fractal dimension to characterize the presence of architectural distortion in mammographic ROIs has been explored recently. Guo *et al.* [145] investigated the characterization of architectural distortion using the Hausdorff dimension, and a support vector machine classifier to distinguish between mammographic ROIs exhibiting architectural distortion and those with normal mammographic patterns. A set of 40 ROIs was selected from the MIAS database [42] (19 ROIs with architectural distortion and 21 ROIs with normal tissue patterns). The authors reported a classification accuracy of 72.5%. Tourassi *et al.* [146] studied the use of fractal dimension to differentiate between normal and architectural distortion patterns in mammographic ROIs. The dataset used in the investigation contained 112 ROIs with architectural distortion patterns, and 1, 388 ROIs exhibiting normal tissue patterns. An area under the ROC curve of $A_z = 0.89$ was obtained. The authors also reported that the average fractal dimension of ROIs exhibiting architectural distortion was observed to be lower than that of ROIs with normal patterns, and that the observed difference was statistically significant under an independent-sample, two-tailed t-test.

Matsubara *et al.* [147] used mathematical morphology to detect architectural distortion around the skin line, and a concentration index to detect architectural distortion within the mammary gland; the authors reported a sensitivity of 94% with 2.3 false positives per image, and 84% with 2.4 false positives per image, respectively. In a later report from the same research group, Ichikawa *et al.* [60] presented a method to detect architectural distortion that encompasses the detection of linear structures using the mean curvature of the image, the computation of a concentration index that indicates the presence of stellate structures over half-circles, and the detection of architectural distortion based on a set of local features that includes the concentration index; the authors reported a sensitivity of 68% with 3.4 false positives per image. Mudigonda and Rangayyan [148] proposed the use of texture flow-field to detect architectural distortion, based on the local coherence of texture orientation; only preliminary results were given, indicating the potential of the technique in the detection of architectural distortion.

Eltonsy *et al.* [149] proposed a method for the detection of masses and architectural distortion based on the identification of points surrounded by concentric layers of image activity. A test dataset of 80 images was used in the evaluation of the technique, containing 13 masses, 38 masses accompanied by architectural distortion, and 29 images exhibiting only architectural distortion. The authors reported an overall sensitivity of 91.3% with 9.1 false positives per image. A sensitivity of 93.1% in the detection of pure architectural distortion was also reported at the same level of false positives per image in the overall dataset, without providing a specificity measure applied only to the cases of pure architectural distortion.

Ayres and Rangayyan [13, 150] studied the characterization of architectural distortion in mammographic ROIs using phase portraits. A database of 106 ROIs extracted from the Mini-MIAS database was used, containing 17 cases of architectural distortion, 45 normals, two ROIs with malignant calcifications, and 44 masses (eight spiculated malignant, four circumscribed malignant, 11 spiculated benign, and 19 circumscribed benign masses) [13]. A sensitivity of 76.5% and a specificity of 76.4% were obtained, with an area under the ROC curve of $A_z = 0.77$ [13]. The authors extended their work to the detection of architectural distortion in full mammograms [56]: preliminary results indicated a sensitivity of 88% at a high false-positive rate (15 false positives per image) with a dataset of 19 images exhibiting architectural distortion. Further developments of the technique included the rejection of CLS associated with strong gradients [14, 151]: a sensitivity of 84% at 7.8 false positives per image was obtained. In a subsequent work, Rangayyan and Ayres [68, 69] included a shape constraint on the phase portrait model. The technique was evaluated on a database of 19 images containing architectural distortion and 41 normal mammograms. A sensitivity of 84% was obtained at 4.5 false positives per image. Details of these methods are presented in Chapter 4.

A.10   ANALYSIS OF PRIOR MAMMOGRAMS

Screening mammography has a limited sensitivity [80], and it has been observed that subtle signs of abnormality can be seen in a significant fraction of previous screening mammograms of screen-detected or interval cases of breast cancer [81], hereafter referred to as prior mammograms. It is possible that such cases of missed signs of abnormality present indistinct, subtle, or hard-to-detect features related to early signs of breast cancer. Based on these observations, a few researchers have analyzed prior mammograms in efforts to improve the detection of early signs of breast cancer.

Sameti *et al.* [154] studied the structural differences between the regions that subsequently developed malignant masses on mammograms, and other normal areas in images taken in the last screening instance prior to the detection of tumors. Manually identified circular ROIs were transformed into their optical-density equivalents, and further divided into three types of regions representing low, medium, and high optical density. Based upon the regions, a set of photometric and texture features was extracted. It was reported that in 72% of the 58 breast cancer cases studied, it was possible to realize the differences between malignant tumor regions and normal tissues in the prior mammograms.

Petrick *et al.* [155] studied the effectiveness of their mass-detection method in the detection of masses in prior mammograms. The dataset used included 92 images (54 malignant and 38 benign) from 37 cases (22 malignant and 15 benign). Their detection methods achieved a "by film" mass-detection sensitivity of 51% with 2.3 false positives per image; a slightly better accuracy of 57% was achieved in detecting only malignant tumors. The detection scheme of Petrick *et al.* attempts to segment salient densities by employing region growing after enhancement of contrast in the image. Such an intensity-based segmentation approach fails to detect the developing densities in prior mammograms due to the inadequate contrast of potentially abnormal regions before the masses are actually formed.

Zheng *et al.* [156] investigated the performance of a CAD algorithm for the detection of masses in current and prior mammograms in two scenarios: when the algorithm was optimized with current mammograms, and when the algorithm was optimized with prior mammograms. The CAD algorithm consisted of three steps: difference-of-Gaussian filtering and thresholding for the initial selection of potential lesion sites; adaptive region growing and topological analysis of the suspicious regions to eliminate false positives; and feature extraction (including shape, histogram, and texture features) and classification using an artificial neural network (ANN). A database of 260 pairs of consecutive mammograms was used in this work, where the latest image showed one or two masses, and the prior image had been originally classified as negative or probably benign. The first two steps of the CAD algorithm were applied to both the current and prior images of the database, producing a set of $1,449$ suspicious ROIs, which were classified according to the true mass location in the current mammograms. The ROIs were classified into the normal and mass categories using the third step of the CAD algorithm (feature extraction and ANN classification). The authors reported that training the ANN with the current mammograms resulted in areas under the ROC curves of 0.89 ± 0.01 and 0.65 ± 0.02 when classifying ROIs from the current and prior mammograms, respectively. When the ANN was trained with ROIs from the prior mammograms, areas under the ROC curve of 0.81 ± 0.02 and 0.71 ± 0.02 were obtained in the classification of ROIs from the current and prior mammograms, respectively. The results indicate the importance of developing CAD algorithms that incorporate knowledge about particular features of early signs, as opposed to the application of methods designed for well-developed masses, for the detection of early signs of breast cancer.

The method described in Section 4.3 has recently been extended for the detection of architectural distortion in prior mammograms [186, 187, 188, 189, 191, 192]. Several features computed from ROIs detected using the node map, including Haralick's texture features, Laws' texture energy measures, fractal dimension, and measures of angular spread of energy in the Fourier domain have led to detection accuracy of the order of 80% at about 6 false positives per image. These recent works demonstrate the potential to detect breast cancer at stages prior to the formation of readily evident masses.

Simultaneous analysis of current and prior mammograms could improve the performance of radiologists in the detection of breast cancer, and may also enhance the performance of CAD systems in the same task. Burnside *et al.* [157] analyzed the impact of the availability of prior mammograms on the clinical outcomes of diagnostic and screening mammography. The authors concluded that comparison with previous mammograms significantly improves the specificity but not the sensitivity of screening mammography, and increases the sensitivity of diagnostic mammography. Ciatto *et al.* [158] compared single, double, and CAD-assisted reading of negative prior (screening) mammograms in interval-cancer cases. It was observed that CAD-assisted reading was almost as sensitive as double reading, and significantly more specific.

A.11   FULL-FIELD DIGITAL MAMMOGRAPHY

Full-field digital mammography, although not a CAD technology in strict terms, has several advantageous features that can be explored by a CAD system. In a digital imaging system, the steps of image acquisition, processing, display, and storage are decoupled, allowing the optimization of each of these procedures. Several authors have presented reviews of technologies for full-field digital mammography, e.g., James [159], Pisano [160], and Yaffe [161]. Lewin *et al.* [162] compared the performance of full-field digital mammography and screen-film mammography for the detection of breast cancer in a screening population. It was observed that no statistically significant difference ($p > 0.1$) existed in cancer detection, and that digital mammography resulted in fewer recalls than screen-film mammography ($p < 0.001$).

A.12   INDEXED ATLASES, DATA MINING, AND CONTENT-BASED RETRIEVAL

The advent of mammographic screening programs has generated a variety of data, such as patient reports and mammographic images, stored in many databases across health centers and universities around the world. Such a wealth of data could be explored through the use of information management technologies, benefiting researchers, clinical practitioners, students, patients, companies engaged in research and development related to CAD systems, and other participants in the effort to reduce breast cancer mortality.

The Merriam-Webster dictionary defines data as "factual information (such as measurements or statistics) used as a basis for reasoning, discussion, or calculation", and information as "knowledge obtained from investigation, study, or instruction". Data represent a meaningless entity that is transformed into information through the process of analysis and attribution of meaning. It is necessary to develop the proper computational tools in order to obtain useful information from the vast amounts of data present in mammographic and associated databases.

One can contemplate upon the usefulness of efficiently retrieving and analyzing information, by observing how search engines help in taming the massive complexity and quantity of data available on the Internet. Nevertheless, it is necessary to tailor the information retrieval tools to the nature of the information being retrieved. Some researchers have investigated the application of content-based image retrieval (CBIR) [163, 164, 165] and data mining [166] techniques to explore the richness present in databases of mammograms and patient information. Honda *et al.* [163] presented a CBIR system based on textural features and PCA: the authors reported a precision rate between 25% and 100%. Nakagawa *et al.* [164] presented a technique for CBIR where mammographic ROIs containing masses were represented by autocorrelation measures. The authors observed that the technique allowed the retrieval of ROIs that were visually similar to a given ROI, used as a query. Nevertheless, the visual similarity did not imply an agreement between the radiologist's assessment of the query ROI and the retrieved ROIs: an agreement of 29% was obtained in the shape of the mass, and 34% in the description of the mass border. Kinoshita *et al.* developed a CBIR system that incorporated measures of shape, size, and texture for the description of the breast area, and a Kohonen self-organizing map for the retrieval of images. A dataset of 1, 080 mammograms was used to test the system, and precision rates between 78% and 83% were obtained. Azevedo-Marques et al. [190] proposed methods to incorporate experts' opinion and guidance in a CBIR system through relevance feedback.

Alto *et al.* [167] investigated the suitability of objective measures of shape, edge sharpness, and texture to retrieve mammograms with masses having similar features. The retrieval precision was determined to be 91% when using the three most-effective features investigated by the authors, namely fractional concavity, acutance, and sum entropy (a texture measure defined by Haralick [118]). Figure A.3 illustrates the process of feature extraction for a macrolobulated benign mass. The ROI of the mass is shown in Figure A.3a. Figure A.3b displays the contour of the mass, with concave parts in black and convex parts in white: fractional concavity is the fractional length of the concave segments to the total length of the contour. Figure A.3c shows the ribbon of pixels at the boundary of the mass which is used to compute texture measures. Figure A.3d exhibits a set of line segments (in black) perpendicular to the contour of the mass: the image intensity along these perpendicular line segments is used to compute acutance, a measure of edge sharpness. The results of the retrieval operation for a malignant tumor are illustrated in Figure A.8.

Indexed atlases can be developed to help in the teaching and training of radiologists, and combined with content-based retrieval tools to help radiologists in the decision-making process for difficult-to-diagnose cases. Alto *et al.* [168, 169] discuss several issues related to an effective design of an indexed atlas of digital mammograms for CAD of breast cancer. In particular, the use of objective measures derived by the application of image processing techniques to represent diagnostic features in mammograms could lead to semantic indexing, data mining, content-based retrieval, and comparative analysis of cases. Guliato *et al.* [170] developed an indexed atlas of digital mammograms which supports content-based retrieval of mammograms for research purposes, and provides a tutorial system for educators based on a formal specification of knowledge (an ontology[193]) related to the clinical analysis of mammograms.

A.13 COMPUTER-AIDED DIAGNOSIS OF BREAST CANCER

Computer-aided diagnosis techniques [171, 172, 173] could offer a cost-effective alternative to double reading as a means of reducing errors. A CAD system could act as a second reader, prompting

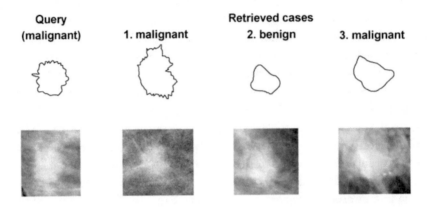

Figure A.8: Content-based retrieval with a microlobulated malignant tumor query sample using the three features fractional concavity (shape), acutance (edge sharpness), and sum entropy (texture measure). In each case, the gray-scale ROI and the corresponding contour drawn by a radiologist are shown. The top three matches selected from the database are shown.

the radiologist to review areas in a mammogram deemed to be suspicious by specialized computer algorithms. A typical CAD session works as follows:

1. The radiologist performs the first reading of the mammogram, recording any questionable or suspicious areas. Optionally, the radiologist could digitally enhance the mammographic image in order to pay closer attention to subtle details that could suggest the presence of lesions.

2. The CAD system scans the mammogram in order to detect suspicious features.

3. The radiologist then analyzes the prompts given by the CAD system to verify whether any suspicious area was left unchecked in the first reading.

CAD algorithms may also be employed to estimate the likelihood that a given lesion is malignant or benign, with such an estimate reviewed subsequently by the radiologist.

The potential benefits of CAD technology motivated the development of several commercial CAD systems, such as the "ImageChecker" (R2 Technology, Sunnyvale, CA [174]) and "Second-Look" (iCAD, Nashua, NH [71]). A recent study has demonstrated that CAD systems can improve a radiologist's sensitivity without a substantial increase in the recall rate [175]. However, another study [176] has indicated that the use of CAD technology resulted in lower specificity, reducing the overall accuracy of interpretation of screening mammograms, while increasing the rate of biopsy.

Ciatto *et al.* [177] compared conventional mammogram reading and CAD reading in a national proficiency test of screening mammography in Italy. The authors concluded that the performance of single reading with CAD is similar to that of double reading. Freer and Ulissey [178] performed a prospective study of the effect of CAD on screening, where 12, 860 screening mammograms were interpreted with the help of a CAD system over a 12-month period. It was observed that the number of cancers detected increased by 19.5%, and the proportion of early-stage malignancies detected increased from 73% to 78%. The recall rate increased from 6.5% to 7.7%, and the positive-predictive value of biopsy remained unchanged at 38%. The study led to the conclusion that CAD can improve the detection of early-stage malignancies without an excessively adverse effect on the recall rate or the positive-predictive value of biopsy.

Burhenne *et al.* [179] studied the performance of a commercial CAD system in the detection of masses and calcifications in screening mammography, obtaining a sensitivity of 75% in the detection of masses and architectural distortion, at one false positive per image. Evans *et al.* [180] investigated the ability of a commercial CAD system to mark invasive lobular carcinoma of the breast: the system identified correctly 17 of 20 cases of architectural distortion. Birdwell *et al.* [181] evaluated the performance of a commercial CAD system in marking cancers that were overlooked by radiologists: the software detected five out of six cases of architectural distortion, and 77% of the previously missed lesions, at 2.9 false positives per image.

However, Baker *et al.* [182] found the sensitivity of two commercial CAD systems to be poor in detecting architectural distortion: fewer than 50% of the 45 cases of architectural distortion presented were detected (with a lower image-based sensitivity of 38%, or 30 out of 80 images, at 0.7 false positives per image). Broeders *et al.* [183] suggested that improvements in the detection of architectural distortion could lead to an effective improvement in the prognosis of breast cancer patients.

These findings indicate the need for further research in this area, and the development of algorithms designed specifically to characterize and detect architectural distortion.

A.14 REMARKS

Mammograms are difficult images to interpret, due to the superposition of various anatomical features in the image, and the subtle nature of signs of breast cancer (especially, signs of early breast cancer). The development of CAD algorithms for breast cancer is an active research field, where the goal is to improve the detection rates of breast cancer. The aforementioned goal is pursued through the development of CAD techniques for the enhancement of mammographic images (to facilitate the perception of signs of breast cancer), detection of signs of breast cancer (such as calcifications, masses, asymmetry, and architectural distortion), and the development of efficient systems for the acquisition and management of mammographic images. The development of indexed atlases and data-mining systems may facilitate the analysis of mammographic images and the training of new radiologists in the reading of mammograms.

Several CAD techniques for the detection of masses and calcifications in mammograms have been developed; proportionately smaller numbers of methods for the detection of architectural distortion and bilateral asymmetry have been proposed. A comprehensive review of the literature of

CAD techniques for breast cancer, presented in this appendix, indicates that further developments are necessary in the detection of subtle signs of early breast cancer.

APPENDIX B

Event detection in medical images

Concepts related to the performance analysis of systems for the detection of events in images are discussed in this Appendix. Such concepts are applicable to the study of systems for the detection of abnormalities in mammographic images, such as architectural distortion. A more comprehensive presentation of the concepts shown in this appendix can ber found in [40, 67].

B.1   SENSITIVITY AND SPECIFICITY

Consider a set of RBIs obtained from mammographic images, where some ROIs exhibit normal tissue patterns (normal ROIs) whereas others display architectural distortion (abnormal ROIs). The true nature of each mammographic ROI, called the "ground truth", is established by means such as expert opinion or biopsy. Normal ROIs will be referred to as "negative findings", or "negatives". Likewise, abnormal ROIs will be labeled as "positive findings", or "positives".

Consider also a computer system (hereafter referred to as the classifier) designed to separate the positives from the negatives. Sensitivity is defined as the fraction of positives that are identified as such by the classifier. The fraction of negatives identified as such by the classifier is called specificity.

In mathematical terms, let us define the following quantities:

- P is the number of positives.

- N is the number of negatives.

- TP is the number of true positives, that is, positives which were categorized as such by the classifier.

- TN is the number of true negatives, that is, negatives which were categorized as such by the classifier.

- FP is the number of false positives, that is, negatives which were categorized as positives by the classifier.

- FN is the number of false negatives, that is, positives which were categorized as negatives by the classifier.

Sensitivity and specificity are defined as:

$$\text{sensitivity} = \frac{TP}{P} = \frac{TP}{TP + FN}$$

and

$$\text{specificity} = \frac{TN}{N} = \frac{TN}{TN + FP}.$$

An ideal classifier would achieve sensitivity and specificity values of 1.0, or 100%.

B.2   RECEIVER OPERATING CHARACTERISTIC ANALYSIS

Let Ω be the set of elements being classified (for instance, mammographic ROIs), and $\omega \in \Omega$ be an element in the set. Several classifiers are comprised of a function $g : \Omega \mapsto \mathbb{R}$ and a decision threshold T such that if $g(\omega) > T$, then the element ω is classified as a positive; otherwise, ω is classified as a negative element. The value $g(\omega)$ is called the discriminant value of ω.

Changing the threshold value T affects the sensitivity and specificity of the classifier. A high value of T implies a rigorous or strict classifier, where few negative elements will be misclassified as positives, but many positive elements may not pass the high threshold criterion. In this case, a high specificity is achieved, at the potential expense of a low sensitivity. At the other extreme, consider the situation where the value of T is low. Then, the classifier will accept more elements as positives, and potentially commit the error of classifying negative elements as positives. Such a classifier would have a high sensitivity, but low specificity.

The receiver operating characteristic (ROC) curve is a graph of all possible values of sensitivity and specificity that are obtained by varying the threshold value T. It is customary to plot the ROC curve with the sensitivity values on the ordinate (y-axis), and the quantity

$$FPF = 1 - \text{specificity},$$

known as the false-positive fraction, on the abscissa (x-axis). The area under the ROC curve is denoted by the symbol A_z, and indicates the overall performance of the classifier. A perfect classifier will have $A_z = 1$, whereas a random classifier (which assigns a random label to each element) will have $A_z = 0.5$.

The problem of detection differs from the problem discussed in the preceding paragraphs (which is the classification problem). In the detection problem, the goal is to identify whether there is an abnormality present in a given mammographic image, and to mark the location of the abnormality, if present. A detection system, in the context of the present work, may mark several locations in the image (detection marks) as potential sites of architectural distortion. In this new context, sensitivity refers to the fraction of images containing architectural distortion in which at least one detection mark corresponds to the true site of architectural distortion. Specificity, on the other hand, is undefined: there is no pre-specified number of negatives. An alternative measure that conveys the idea of specificity is the false alarm rate of the system, specified by the number of false positives (or false detections) per image.

Similar to the classification problem, a detection system may provide a discriminant value for each tentative detection mark, and a threshold value may determine the acceptance or rejection of a detection mark. Changing the threshold will change the sensitivity, as well as the rate of false positives per image. A graph that displays the relationship between the sensitivity and the number of false positives per image is called the free-response ROC curve, abbreviated as FROC.

Bibliography

[1] C.-F. Yang, C. M. Crosby, A. R. K. Eusufzai, and R. E. Mark. Determination of paper sheet fiber orientation distributions by a laser optical diffraction method. *Journal of Applied Polymer Science*, 34(3):1145–1157, 1987. DOI: 10.1002/app.1987.070340323 1

[2] F. Thorarinsson, S. G. Magnusson, and A. Bjornsson. Directional spectral analysis and filtering of geophysical maps. *Geophysics*, 53(12):1587–1591, 1988. DOI: 10.1190/1.1442440 1

[3] V. Carrere. Development of multiple source data processing for structural analysis at a regional scale. *Photogrammetric Engineering and Remote Sensing*, 56(5):587–595, 1990. 1

[4] P. Embree and J. P. Burg. Wide-band velocity filtering — the pie slice process. *Geophysics*, 28:948–974, 1963. DOI: 10.1190/1.1439310 1, 2

[5] S. Treitel, J. L. Shanks, and C. W. Frasier. Some aspects of fan filtering. *Geophysics*, 32:789–806, 1967. DOI: 10.1190/1.1439889 1, 2

[6] V. Bezvoda, J. Ježek, and K. Segeth. FREDPACK- A program package for linear filtering in the frequency domain. *Computers & Geosciences*, 16(8):1123–1154, 1990. DOI: 10.1016/0098-3004(90)90053-V 1, 2

[7] S. Shah and P. S. Sastry. Fingerprint classification using a feedback-based line detector. *IEEE Transactions on Systems, Man, and Cybernetics - part B: Cybernetics*, 34(1):85–94, February 2004. DOI: 10.1109/TSMCB.2002.806486 1

[8] C. Frank, B. MacFarlane, P. Edwards, R. Rangayyan, Z. Q. Liu, S. Walsh, and R. Bray. A quantitative analysis of matrix alignment in ligament scars: A comparison of movement versus immobilization in an immature rabbit model. *Journal of Orthopaedic Research*, 9(2):219–227, 1991. DOI: 10.1002/jor.1100090210 1

[9] S. Chaudhuri, H. Nguyen, R. M. Rangayyan, S. Walsh, and C. B. Frank. A Fourier domain directional filtering method for analysis of collagen alignment in ligaments. *IEEE Transactions on Biomedical Engineering*, 34(7):509–518, 1987. DOI: 10.1109/TBME.1987.325980 1, 2

[10] K. Eng, R. M. Rangayyan, R. C. Bray, C. B. Frank, L. Anscomb, and P. Veale. Quantitative analysis of the fine vascular anatomy of articular ligaments. *IEEE Transactions on Biomedical Engineering*, 39(3):296–306, 1992. DOI: 10.1109/10.125015 1

[11] M. Sonka, W. Park, and E. A. Hoffman. Rule-based detection of intrathoracic airway trees. *IEEE Transactions on Medical Imaging*, 15(3):314–326, June 1996. DOI: 10.1109/42.500140 1

[12] D. Aykac, E. A. Hoffman, G. McLennan, and J. M. Reinhardt. Segmentation and analysis of the human airway tree from three-dimensional x-ray CT images. *IEEE Transactions on Medical Imaging*, 22(8):940–950, August 2003. DOI: 10.1109/TMI.2003.815905 1

[13] F. J. Ayres and R. M. Rangayyan. Characterization of architectural distortion in mammograms. *IEEE Engineering in Medicine and Biology Magazine*, 24(1):59–67, January 2005. DOI: 10.1109/MEMB.2005.1384102 1, 116

[14] R. M. Rangayyan and F. J. Ayres. Gabor filters and phase portraits for the detection of architectural distortion in mammograms. *Medical and Biological Engineering and Computing*, 44:883–894, August 2006. DOI: 10.1007/s11517-006-0088-3 1, 2, 7, 26, 68, 116

[15] N. Karssemeijer and G. M. te Brake. Detection of stellate distortions in mammograms. *IEEE Transactions on Medical Imaging*, 15(5):611–619, 1996. DOI: 10.1109/42.538938 1, 3, 75, 106

[16] R. Zwiggelaar, S. M. Astley, C. R. M. Boggis, and C. J. Taylor. Linear structures in mammographic images: Detection and classification. *IEEE Transactions on Medical Imaging*, 23(9):1077–1086, September 2004. DOI: 10.1109/TMI.2004.828675 1, 10, 75, 112

[17] A. D. Hoover, V. Kouznetsova, and M. Goldbaum. Locating blood vessels in retinal images by piecewise threshold probing of a matched filter response. *IEEE Transactions on Medical Imaging*, 19(3):203–210, March 2000. DOI: 10.1109/42.845178 1

[18] M. P. Saren, R. Serimaa, and Y. Tolonen. Determination of fiber orientation in Norway spruce using x-ray diffraction and laser scattering. *Holz als Roh-und Werkstoff*, 64(3):183–188, June 2006. DOI: 10.1007/s00107-005-0076-6 1

[19] N. Dalal, B. Triggs, and C. Schmid. Human detection using oriented histograms of flow and appearance. In *Proceedings of the 9th European Conference on Computer Vision - ECCV 2006*, pages 428–441, Graz, Austria, May 2006. DOI: 10.1007/11744047_33 1

[20] M. Moganti, F. Ercal, C. H. Dagli, and S. Tsunekawa. Automatic PCB inspection algorithms: A survey. *Computer Vision and Image Understanding*, 63(2):287–313, March 1996. DOI: 10.1006/cviu.1996.0020 1

[21] R. Plamondon and S. N. Srihari. Online and off-line handwriting recognition: a comprehensive survey. *IEEE Transactions on Pattern Analysis and Machine Intelligence*, 22(1):63–84, January 2000. DOI: 10.1109/34.824821 1

[22] R. N. Dixon and C. J. Taylor. Automated asbestos fibre counting. *Institute of Physics Conference Series*, 44:178–185, 1979. 2, 10, 25, 112

[23] E. R. Davies, M. Bateman, D. R. Mason, J. Chambers, and C. Ridgway. Design of efficient line segment detectors for cereal grain inspection. *Pattern Recognition Letters*, 24:413–428, 2003. DOI: 10.1016/S0167-8655(02)00266-0 2

[24] P. J. B. Hancock, R. J. Baddeley, and L. S. Smith. The principal components of natural images. *Network: Computation in Neural Systems*, 3(1):61–70, February 1992. DOI: 10.1088/0954-898X/3/1/008 2

[25] G. Heidemann. The principal components of natural images revisited. *IEEE Transactions on Pattern Analysis and Machine Intelligence*, 28(5):822–826, May 2006. DOI: 10.1109/TPAMI.2006.107 2

[26] W. T. Freeman and E. H. Adelson. The design and use of steerable filters. *IEEE Transactions on Pattern Analysis and Machine Intelligence*, 13(9):891–906, September 1991. DOI: 10.1109/34.93808 2, 3, 5, 10, 112

[27] D. Gabor. Theory of communication. *Journal of the Institute of Electrical Engineers*, 93:429–457, 1946. 2, 7

[28] C. K. Chui. *An Introduction to Wavelets*, volume 1 of *Wavelet Analysis and its Applications*. Academic Press, San Diego, CA, 1992. 2, 7

[29] B. S. Manjunath and W. Y. Ma. Texture features for browsing and retrieval of image data. *IEEE Transactions on Pattern Analysis and Machine Intelligence*, 18(8):837–842, 1996. DOI: 10.1109/34.531803 2, 7

[30] R. J. Ferrari, R. M. Rangayyan, J. E. L. Desautels, and A. F. Frère. Analysis of asymmetry in mammograms via directional filtering with Gabor wavelets. *IEEE Transactions on Medical Imaging*, 20(9):953–964, 2001. DOI: 10.1109/42.952732 1, 2, 7, 26, 112

[31] L. T. Bruton and N. R. Bartley. Using nonessential singularities of the second kind in two-dimensional filter design. *IEEE Transactions on Circuits and Systems*, 36(1):113–116, 1989. DOI: 10.1109/31.16572 2

[32] N. Merlet and J. Zerubia. New prospects in line detection by dynamic programming. *IEEE Transactions on Pattern Analysis and Machine Intelligence*, 18(4):426–431, April 1996. DOI: 10.1109/34.491623 2

[33] D. S. Guru, B. H. Shekar, and P. Nagabhushan. A simple and robust line detection algorithm based on small eigenvalue analysis. *Pattern Recognition Letters*, 25:1–13, 2003. DOI: 10.1016/j.patrec.2003.08.007 2

[34] F. J. Ayres and R. M. Rangayyan. Design and performance analysis of oriented feature detectors. *Journal of Electronic Imaging*, 16(2):12 pages, April 2007. article number 023007. DOI: 10.1117/1.2728751 2, 11

[35] F. J. Ayres and R. M. Rangayyan. Performance analysis of oriented feature detectors. In *Proceedings of SIBGRAPI 2005: XVIII Brazilian Symposium on Computer Graphics and Image Processing*, pages 147–154, Natal, Brazil, October 2005. IEEE Computer Society Press. DOI: 10.1109/SIBGRAPI.2005.38 2

[36] R. M. Rangayyan. *Biomedical Image Analysis*. The Biomedical Engineering Series. CRC Press, Boca Raton, FL, 2005. 7, 36, 99, 100, 101

[37] R. Zwiggelaar, T. C. Parr, and C. J. Taylor. Finding orientated line patterns in digital mammographic images. In *Proceedings of the 7th British Machine Vision Conference*, pages 715–724, Edinburgh, UK, 1996. 10, 112

[38] T. Lindeberg. Edge detection and ridge detection with automatic scale selection. *International Journal of Computer Vision*, 30(2):117–154, 1998. DOI: 10.1023/A:1008045108935 10, 112

[39] P. J. Burt and E. H. Adelson. The Laplacian pyramid as a compact image code. *IEEE Transactions on Communications*, 31(4):532–540, April 1983. DOI: 10.1109/TCOM.1983.1095851 10, 11

[40] C. E. Metz. Basic principles of ROC analysis. *Seminars in Nuclear Medicine*, VIII(4):283–298, 1978. DOI: 10.1016/S0001-2998(78)80014-2 15, 123

[41] N. Cerneaz and M. Brady. Finding curvilinear structures in mammograms. In *CVRMed '95: Proceedings of the First International Conference on Computer Vision, Virtual Reality and Robotics in Medicine*, pages 372–382, London, UK, 1995. Springer-Verlag. DOI: 10.1007/BFb0034973 25

[42] J. Suckling, J. Parker, D. R. Dance, S. Astley, I. Hutt, C. R. M. Boggis, I. Ricketts, E. Stamakis, N. Cerneaz, S.-L. Kok, P. Taylor, D. Betal, and J. Savage. The Mammographic Image Analysis Society Digital Mammogram Database. In A. G. Gale, S. M. Astley, D. D. Dance, and A. Y. Cairns, editors, *Digital Mammography: Proceedings of the 2nd International Workshop on Digital Mammography*, pages 375–378, York, UK, July 1994. Elsevier. 25, 68, 80, 91, 103, 106, 107, 108, 115

[43] D. Scutt, G. A. Lancaster, and J. T. Manning. Breast asymmetry and predisposition to breast cancer. *Breast Cancer Research*, 8:R14, DOI: 10.1186/bcr1388, 2006. DOI: 10.1186/bcr1388 26, 112

[44] A. R. Rao and R. C. Jain. Computerized flow field analysis: Oriented texture fields. *IEEE Transactions on Pattern Analysis and Machine Intelligence*, 14(7):693–709, July 1992. DOI: 10.1109/34.142908 31, 34, 36, 41, 43, 47

[45] A. R. Rao. *A Taxonomy for Texture Description and Identification*. Springer-Verlag, New York, NY, 1990. 31, 43, 67

[46] C. R. Wylie and L. C. Barrett. *Advanced Engineering Mathematics*. McGraw-Hill, New York, NY, 6th edition, 1995. 31, 32

[47] E. Kreyszig. *Advanced Engineering Mathematics*. Wiley, New York, NY, 1983. 37

[48] C.-F. Shu and R. C. Jain. Vector field analysis for oriented patterns. *IEEE Transactions on Pattern Analysis and Machine Intelligence*, 16(9):946–950, September 1994. DOI: 10.1109/34.310692 43, 45, 47

[49] F. J. Ayres and R. M. Rangayyan. An iterative linear algorithm for the analysis of oriented patterns. In E. R. Dougherty, J. T. Astola, and K. O. Egiazarian, editors, *Electronic Imaging 2004*, volume 5298, pages 232–241, San Jose, CA, 2004. SPIE. 43, 45, 46

[50] D. W. Marquardt. An algorithm for the least-squares estimation of nonlinear parameters. *Journal of the Society for Industrial and Applied Mathematics*, 11:431–441, 1963. DOI: 10.1137/0111030 43, 48

[51] F. J. Ayres and R. M. Rangayyan. Optimization procedures for the estimation of phase portrait parameters of orientation fields. In E. R. Dougherty, J. T. Astola, K. O. Egiazarian, N. M. Nasrabadi, and S. A. Rizvi, editors, *Electronic Imaging 2006*, volume 6064, San Jose, CA, 2006. SPIE. 43

[52] S. Kirkpatrick, C. D. Gelatt, and M. P. Vecchi. Optimization by simulated annealing. *Science*, 220(4598):671–680, 1983. DOI: 10.1126/science.220.4598.671 43, 48, 79

[53] J. Kennedy and R. Eberhart. Particle swarm optimization. In *Proceedings of the IEEE International Conference on Neural Networks*, pages 1942–1948, Perth, Australia, November 1995. DOI: 10.1109/ICNN.1995.488968 43, 48

[54] M. Galassi, J. Davies, J. Theiler, B. Gough, G. Jungman, M. Booth, and F. Rossi. *GNU Scientific Library: Reference Manual*. Network Theory, Bristol, UK, 2nd edition, 2003. 48

[55] M. Clerc and J. Kennedy. The particle swarm - explosion, stability, and convergence in a multidimensional complex space. *IEEE Transactions on Evolutionary Computation*, 6(1):58–73, February 2002. DOI: 10.1109/4235.985692 48, 50

[56] F. J. Ayres and R. M. Rangayyan. Detection of architectural distortion in mammograms using phase portraits. In J. M. Fitzpatrick and M. Sonka, editors, *Proceedings of SPIE Medical Imaging 2004: Image Processing*, volume 5370, pages 587–597, San Diego, CA, February 2004. 61, 116

[57] E. R. Dougherty. *An Introduction to Morphological Image Processing*. SPIE Press, 1992. 67, 80

[58] F. J. Ayres and R. M. Rangayyan. Detection of architectural distortion in mammograms via analysis of phase portraits and curvilinear structures. In J. Hozman and P. Kneppo, editors, *Proceedings of EMBEC'05: 3rd European Medical & Biological Engineering Conference*, volume 11, pages 1768–1773, Prague, Czech Republic, November 2005. 68, 75

[59] R. J. Ferrari, R. M. Rangayyan, R. A. Borges, and A. F. Frère. Segmentation of the fibro-glandular disc in mammograms using Gaussian mixture modeling. *Medical and Biological Engineering and Computing*, 42:378–387, 2004. DOI: 10.1007/BF02344714 75, 103

[60] T. Ichikawa, T. Matsubara, T. Hara, H. Fujita, T. Endo, and T. Iwase. Automated detection method for architectural distortion areas on mammograms based on morphological processing and surface analysis. In J. M. Fitzpatrick and M. Sonka, editors, *Proceedings of SPIE Medical Imaging 2004: Image Processing*, pages 920–925, San Diego, CA, February 2004. SPIE. 75, 115

[61] N. Cerneaz and M. Brady. Enriching digital mammogram image analysis with a description of the curvi-linear structures. In A. G. Gale, S. M. Astley, D. R. Dance, and A. Y. Cairns, editors, *Digital Mammography: Proceedings of the 2nd International Workshop on Digital Mammography*, pages 297–306, York, UK, July 1994. Elsevier. 75

[62] N. Otsu. A threshold selection method from gray-level histograms. *IEEE Transactions on Systems, Man, and Cybernetics*, 9(1):62–66, 1979. DOI: 10.1109/TSMC.1979.4310076 75

[63] R. J. Ferrari, R. M. Rangayyan, J. E. L. Desautels, R. A. Borges, and A. F. Frère. Identification of the breast boundary in mammograms using active contour models. *Medical and Biological Engineering and Computing*, 42:201–208, 2004. DOI: 10.1007/BF02344632 1, 75, 103

[64] M. Sonka, V. Hlavac, and R. Boyle. *Image Processing, Analysis and Machine Vision*. Chapman & Hall, London, UK, 1st edition, 1993. 75

[65] J. Canny. A computational approach to edge detection. *IEEE Transactions on Pattern Analysis and Machine Intelligence*, 8(6):679–698, 1986. DOI: 10.1109/TPAMI.1986.4767851 75

[66] N. Gershenfeld. *The Nature of Mathematical Modeling*. Cambridge University Press, Cambridge, UK, 1999. 79

[67] C. E. Metz. ROC methodology in radiologic imaging. *Investigative Radiology*, 21:720–733, 1986. DOI: 10.1097/00004424-198609000-00009 80, 123

[68] R. M. Rangayyan and F. J. Ayres. Detection of architectural distortion in mammograms using a shape-constrained phase portrait model. In H. U. Lemke, K. Inamura, K. Doi, M. W. Vannier, and A. G. Farman, editors, *Proceedings of the 20th International Congress and Exhibition on Computer Assisted Radiology and Surgery (CARS 2006)*, pages 334–336, Osaka, Japan, July 2006. Springer. 86, 116

[69] F. J. Ayres and R. M. Rangayyan. Reduction of false positives in the detection of architectural distortion in mammograms by using a geometrically constrained phase portrait model. *International Journal of Computer-Assisted Radiology and Surgery*, 1(6):361–369, April 2007. DOI: 10.1007/s11548-007-0072-x 86, 116

[70] K. M. Abadir and J. R. Magnus. *Matrix Algebra*. Cambridge University Press, New York, NY, 2005. 87

[71] iCAD website, http://www.icadmed.com/, accessed on March 6, 2005. 95, 99, 120

[72] National Cancer Institute of Canada. Canadian cancer statistics 2006, 2006. Available at http://www.cancer.ca/vgn/images/portal/cit_86751114/31/21/ 935505792cw_2006stats_en.pdf.pdf, accessed on June 1st, 2006. 97

[73] A. Jemal, L. X. Clegg, E. Ward, L. A. G. Ries, X. Wu, P. M. Jamison, P. A. Wingo, H. L. Howe, R. N. Anderson, and B. K. Edwards. Annual report to the nation on the status of cancer, 1975-2001, with a special feature regarding survival. *Cancer*, 101(1):3–27, 2004. DOI: 10.1002/cncr.20288 97

[74] C. Di Maggio. State of the art of current modalities for the diagnosis of breast lesions. *European Journal on Nuclear Medicine and Molecular Imaging*, 31, supplement 1:S56–S69, June 2004. DOI: 10.1007/s00259-004-1527-8 97

[75] A. K. Hackshaw and E. A. Paul. Breast self-examination and death from breast cancer: a meta-analysis. *British Journal of Cancer*, 88:1047–1053, 2003. DOI: 10.1038/sj.bjc.6600847 97

[76] M. J. Homer. *Mammographic Interpretation: A Practical Approach*. McGraw-Hill, New York, NY, 2nd edition, 1997. 61, 97

[77] M. A. Schneider. Better detection: Improving our chances. In M. J. Yaffe, editor, *Digital Mammography: 5th International Workshop on Digital Mammography*, pages 3–6, Toronto, ON, Canada, June 2000. Medical Physics Publishing. DOI: 10.1148/radiol.2291032535 97

[78] S. H. Heywang-Köbrunner, I. Schreer, and D. D. Dershaw. *Diagnostic Breast Imaging: Mammography, Sonography, Magnetic Resonance Imaging, and Interventional Procedures*. Thieme Medical Publishers, New York, NY, 1997. 97

[79] B. Cady and M. Chung. Mammographic screening: No longer controversial. *American Journal of Clinical Oncology*, 28(1):1–4, February 2005. DOI: 10.1097/01.coc.0000150720.15450.05 97

[80] R. E. Bird, T. W. Wallace, and B. C. Yankaskas. Analysis of cancers missed at screening mammography. *Radiology*, 184(3):613–617, 1992. 97, 116

[81] J. A. A. M. van Dijck, A. L. M. Verbeek, J. H. C. L. Hendriks, and R. Holland. The current detectability of breast cancer in a mammographic screening program. *Cancer*, 72(6):1933–1938, 1993.
DOI: 10.1002/1097-0142(19930915)72:6%3C1933::AID-CNCR2820720623 97, 116

[82] R. G. Blanks, M. G. Wallis, and S. M. Moss. A comparison of cancer detection rates achieved by breast cancer screening programmes by number of readers, for one and two view mammography: Results from the UK National Health Service Breast Screening Programme. *Journal of Medical Screening*, 5(4):195–201, 1998. DOI: 10.1136/jms.5.4.195 98

[83] G. Cardenosa. *Breast Imaging Companion*. Lippincott-Raven, Philadelphia, PA, 1997. 98

[84] American College of Radiology (ACR). *Illustrated Breast Imaging Reporting and Data System (BI-RADS)*. American College of Radiology, Reston, VA, 3rd edition, 1998. 61, 98

[85] H.-O. Peitgen, editor. *Proceedings of the 6th International Workshop on Digital Mammography*, Bremen, Germany, June 2002. Springer-Verlag. 99

[86] E. Pisano, editor. *Proceedings of the 7th International Workshop on Digital Mammography*, Durham, NC, June 2004. 99

[87] W. M. Morrow, R. B. Paranjape, R. M. Rangayyan, and J. E. L. Desautels. Region-based contrast enhancement of mammograms. *IEEE Transactions on Medical Imaging*, 11(3):392–406, 1992. DOI: 10.1109/42.158944 100, 101

[88] R. M. Rangayyan, L. Shen, Y. Shen, J. E. L. Desautels, H. Bryant, T. J. Terry, N. Horeczko, and M. S. Rose. Improvement of sensitivity of breast cancer diagnosis with adaptive neighborhood contrast enhancement of mammograms. *IEEE Transactions on Information Technology in Biomedicine*, 1(3):161–170, 1997. DOI: 10.1109/4233.654859 100, 101

[89] G. Ram. Optimization of ionizing radiation usage in medical imaging by means of image enhancement techniques. *Medical Physics*, 9(5):733–737, 1982. DOI: 10.1118/1.595119 100

[90] H. P. Chan, C. J. Vyborny, H. MacMahon, C. E. Metz, K. Doi, and E. A. Sickles. ROC studies of the effects of pixel size and unsharp-mask filtering on the detection of subtle microcalcifications. *Investigative Radiology*, 22:581–589, 1987.
DOI: 10.1097/00004424-198707000-00010 100

[91] A. F. Laine, S. Schuler, J. Fan, and W. Huda. Mammographic feature enhancement by multiscale analysis. *IEEE Transactions on Medical Imaging*, 13(4):725–740, December 1994.
DOI: 10.1109/42.363095 100

[92] R. Gordon and R. M. Rangayyan. Feature enhancement of film mammograms using fixed and adaptive neighborhoods. *Applied Optics*, 23(4):560–564, February 1984.
DOI: 10.1364/AO.23.000560 100

[93] R. M. Rangayyan and H. N. Nguyen. Pixel-independent image processing techniques for noise removal and feature enhancement. In *IEEE Pacific Rim Conference on Communications, Computers, and Signal Processing*, pages 81–84, Vancouver, BC, Canada, June 1987. IEEE. 100

[94] A. P. Dhawan, G. Buelloni, and R. Gordon. Enhancement of mammographic features by optimal adaptive neighborhood image processing. *IEEE Transactions on Medical Imaging*, 5(1):8–15, 1986. DOI: 10.1109/TMI.1986.4307733 100

[95] A. P. Dhawan and E. Le Royer. Mammographic feature enhancement by computerized image processing. *Computer Methods and Programs in Biomedicine*, 27:23–35, 1988. DOI: 10.1016/0169-2607(88)90100-9 101

[96] C. Kimme-Smith, R. H. Gold, L. W. Bassett, L. Gormley, and C. Morioka. Diagnosis of breast calcifications: Comparison of contact, magnified, and television-enhanced images. *American Journal of Roentgenology*, 153:963–967, 1989. 101

[97] R. M. Rangayyan, L. Shen, Y. Shen, M. S. Rose, J. E. L. Desautels, H. E. Bryant, T. J. Terry, and N. Horeczko. Region-based contrast enhancement. In R. N. Strickland, editor, *Image-Processing Techniques for Tumor Detection*, pages 213–242. Marcel Dekker, New York, NY, 2002. 101

[98] R. Sivaramakrishna, N. A. Obuchowski, W. A. Chilcote, G. Cardenosa, and K. A. Powell. Comparing the performance of mammographic enhancement algorithms – a preference study. *American Journal of Roentgenology*, 175:45–51, 2000. 101

[99] T.-L. Ji, M. K. Sundareshan, and H. Roehrig. Adaptive image contrast enhancement based on human visual properties. *IEEE Transactions on Medical Imaging*, 13(4):573–586, 1994. DOI: 10.1109/42.363111 101

[100] S. M. Pizer, E. P. Amburn, J. D. Austin, R. Cromartie, A. Geselowitz, T. Geer, B. ter Haar Romeny, J. B. Zimmerman, and K. Zuiderveld. Adaptive histogram equalization and its variations. *Computer Vision, Graphics, and Image Processing*, 39:355–368, 1987. DOI: 10.1016/S0734-189X(87)80186-X 101

[101] A. Laine, J. Fan, and W. H. Yan. Wavelets for contrast enhancement of digital mammography. *IEEE Engineering in Medicine and Biology Magazine*, 14(5):536–550, September/October 1995. DOI: 10.1109/51.464770 101

[102] N. F. Boyd, J. W. Byng, R. A. Jong, E. K. Fishell, L. E. Little, A. B. Miller, G. A. Lockwood, D. L. Tritchler, and M. J. Yaffe. Quantitative classification of mammographic densities and breast cancer risk: results from the Canadian National Breast Screening Study. *Journal of the National Cancer Institute*, 87(9):670–675, May 1995. DOI: 10.1093/jnci/87.9.670 101

[103] J. W. Byng, N. F. Boyd, E. Fishell, R. A. Jong, and M. J. Yaffe. Automated analysis of mammographic densities. *Physics in Medicine and Biology*, 41:909–923, 1996. DOI: 10.1088/0031-9155/41/5/007 101

[104] S. Caulkin, S. Astley, J. Asquith, and C. Boggis. Sites of occurrence of malignancies in mammograms. In N. Karssemeijer, M. Thijssen, J. Hendriks, and L. van Erning, editors, *Proceedings of the 4th International Workshop on Digital Mammography*, pages 279–282, Nijmegen, The Netherlands, June 1998. 102

[105] N. Karssemeijer. Automated classification of parenchymal patterns in mammograms. *Physics in Medicine and Biology*, 43(2):365–378, 1998. DOI: 10.1088/0031-9155/43/2/011 102

[106] R. J. Ferrari, R. M. Rangayyan, J. E. L. Desautels, R. A. Borges, and A. F. Frère. Automatic identification of the pectoral muscle in mammograms. *IEEE Transactions on Medical Imaging*, 23:232–245, 2004. DOI: 10.1109/TMI.2003.823062 102, 103

[107] P. K. Saha, J. K. Udupa, E. F. Conant, D. P. Chakraborty, and D. Sullivan. Breast tissue density quantification via digitized mammograms. *IEEE Transactions on Medical Imaging*, 20(8):792–803, 2001. DOI: 10.1109/42.938247 103

[108] H. D. Cheng, X. Cai, X. Chen, L. Hu, and X. Lou. Computer-aided detection and classification of microcalcifications in mammograms: a survey. *Pattern Recognition*, 36:2967–2991, 2003. DOI: 10.1016/S0031-3203(03)00192-4 104

[109] L. Shen, R. M. Rangayyan, and J. E. L. Desautels. Detection and classification of mammographic calcifications. *International Journal of Pattern Recognition and Artificial Intelligence*, 7(6):1403–1416, 1993. DOI: 10.1142/S0218001493000686 104, 105

[110] L. Shen, R. M. Rangayyan, and J. E. L. Desautels. Application of shape analysis to mammographic calcifications. *IEEE Transactions on Medical Imaging*, 13(2):263–274, June 1994. DOI: 10.1109/42.293919 104, 107

[111] I. N. Bankman, T. Nizialek, I. Simon, O. B. Gatewood, I. N. Weinberg, and W. R. Broody. Segmentation algorithms for detecting microcalcifications in mammograms. *IEEE Transactions on Information Technology in Biomedicine*, 1(2):141–149, June 1997. DOI: 10.1109/4233.640656 104

[112] R. N. Strickland. Wavelet transforms for detecting microcalcifications in mammograms. *IEEE Transactions on Medical Imaging*, 15(2):218–229, 1996. DOI: 10.1109/42.491423 104

[113] I. El-Naqa, Y. Yang, M. N. Wernick, N. P. Galatsanos, and R. M. Nishikawa. A support vector machine approach for detection of microcalcifications. *IEEE Transactions on Medical Imaging*, 21(12):1552–1563, 2002. DOI: 10.1109/TMI.2002.806569 104

[114] L. Wei, Y. Yang, R. M. Nishikawa, M. N. Vernick, and A. Edwards. Relevance vector machine for automatic detection of clustered microcalcifications. *IEEE Transactions on Medical Imaging*, 24(10):1278–1285, 2005. DOI: 10.1109/TMI.2005.855435 104

[115] S.-N. Yu, K.-Y. Li, and Y.-K. Huang. Detection of microcalcifications in digital mammograms using wavelet filter and Markov random field model. *Computerized Medical Imaging and Graphics*, 30:163–173, 2006. DOI: 10.1016/j.compmedimag.2006.03.002 104

[116] S. Yu and L. Guan. A CAD system for the automatic detection of clustered microcalcifications in digitized mammogram films. *IEEE Transactions on Medical Imaging*, 19(2):115–126, February 2000. DOI: 10.1109/42.836371 105

[117] H. Soltanian-Zadeh, F. Rafiee-Rad, and S. Pourabdollah-Nejad. Comparison of multiwavelet, wavelet, Haralick, and shape features for microcalcification classification in mammograms. *Pattern Recognition*, 37:1973–1986, 2004. DOI: 10.1016/j.patcog.2003.03.001 105

[118] R. M. Haralick. Statistical and structural approaches to texture. *Proceedings of the IEEE*, 67:786–804, May 1979. DOI: 10.1109/PROC.1979.11328 105, 107, 119

[119] C. Serrano, J. D. Trujillo, B. Acha, and R. M. Rangayyan. Use of 2D linear prediction error to detect microcalcifications in mammograms. In *CDROM Proceedings of the II Latin American Congress on Biomedical Engineering*, Havana, Cuba, 23-25 May 2001. 105

[120] B. Acha, C. Serrano, R. M. Rangayyan, and J. E. L. Desautels. Detection of microcalcifications in mammograms. In J. S. Suri and R. M. Rangayyan, editors, *Recent Advances in Breast Imaging, Mammography, and Computer-Aided Diagnosis of Breast Cancer*, chapter 9, pages 291–314. SPIE, Bellingham, WA, 2006. 105

[121] G. R. Kuduvalli and R. M. Rangayyan. Performance analysis of reversible image compression techniques for high-resolution digital teleradiology. *IEEE Transactions on Medical Imaging*, 11(3):430–445, September 1992. DOI: 10.1109/42.158947 105

[122] D. Brzakovic, X. M. Luo, and P. Brzakovic. An approach to automated detection of tumors in mammograms. *IEEE Transactions on Medical Imaging*, 9(3):233–241, September 1990. DOI: 10.1109/42.57760 106

[123] W. P. Kegelmeyer, Jr., J. M. Pruneda, P. D. Bourland, A. Hillis, M. W. Riggs, and M. L. Nipper. Computer-aided mammographic screening for spiculated lesions. *Radiology*, 191(2):331–337, 1994. 106

[124] K. I. Laws. Rapid texture identification. In *Proceedings of SPIE Vol. 238: Image Processing for Missile Guidance*, pages 376–380, San Diego, CA, 1980. 106

[125] G. M. te Brake and N. Karssemeijer. Single and multiscale detection of masses in digital mammograms. *IEEE Transactions on Medical Imaging*, 18(7):628–639, July 1999. DOI: 10.1109/42.790462 106

[126] R. M. Rangayyan, N. R. Mudigonda, and J. E. L. Desautels. Boundary modelling and shape analysis methods for classification of mammographic masses. *Medical and Biological Engineering and Computing*, 38:487–496, 2000. DOI: 10.1007/BF02345742 107

[127] B. S. Sahiner, H. P. Chan, N. Petrick, M. A. Helvie, and L. M. Hadjiiski. Improvement of mammographic mass characterization using spiculation measures and morphological features. *Medical Physics*, 28(7):1455–1465, 2001. DOI: 10.1118/1.1381548 107

[128] B. S. Sahiner, H. P. Chan, N. Petrick, M. A. Helvie, and M. M. Goodsitt. Computerized characterization of masses on mammograms: The rubber band straightening transform and texture analysis. *Medical Physics*, 25(4):516–526, 1998. DOI: 10.1118/1.598228 107

[129] N. R. Mudigonda, R. M. Rangayyan, and J. E. L. Desautels. Gradient and texture analysis for the classification of mammographic masses. *IEEE Transactions on Medical Imaging*, 19(10):1032–1043, 2000. DOI: 10.1109/42.887618 107

[130] N. R. Mudigonda, R. M. Rangayyan, and J. E. L. Desautels. Detection of breast masses in mammograms by density slicing and texture flow-field analysis. *IEEE Transactions on Medical Imaging*, 20(12):1215–1227, 2001. DOI: 10.1109/42.974917 107

[131] R. M. Rangayyan, N. M. El-Faramawy, J. E. L. Desautels, and O. A. Alim. Measures of acutance and shape for classification of breast tumors. *IEEE Transactions on Medical Imaging*, 16(6):799–810, December 1997. DOI: 10.1109/42.650876 107

[132] L. Li, R. A. Clark, and J. A. Thomas. Computer-aided diagnosis of masses with full-field digital mammography. *Academic Radiology*, 9:4–12, 2002. DOI: 10.1016/S1076-6332(03)80290-8 108

[133] L. Zheng and A. K. Chan. An artificial intelligent algorithm for tumor detection in screening mammogram. *IEEE Transactions on Medical Imaging*, 20(7):559–567, July 2001. DOI: 10.1109/42.932741 108

[134] S. Liu, C. F. Babbs, and E. J. Delp. Multiresolution detection of spiculated lesions in digital mammograms. *IEEE Transactions on Image Processing*, 10(6):874–884, June 2001. DOI: 10.1109/83.923284 108

[135] R. Zwiggelaar, T. C. Parr, I. W. Hutt, C. J. Taylor, S. M. Astley, and C. R. M. Boggis. Model-based detection of spiculated lesions in mammograms. *Medical Image Analysis*, 3(1):39–63, 1999. DOI: 10.1016/S1361-8415(99)80016-4 108

[136] D. Guliato, R. M. Rangayyan, W. A. Carnielli, J. A. Zuffo, and J. E. L. Desautels. Segmentation of breast tumors in mammograms using fuzzy sets. *Journal of Electronic Imaging*, 12(3):369–378, July 2003. DOI: 10.1117/1.1579017 108

[137] D. Guliato, R. M. Rangayyan, W. A. Carnielli, J. A. Zuffo, and J. E. L. Desautels. Fuzzy fusion operators to combine results of complementary medical image segmentation techniques. *Journal of Electronic Imaging*, 12(3):379–389, July 2003. DOI: 10.1117/1.1578639 108

[138] C. Evans, K. Yates, and M. Brady. Statistical characterization of normal curvilinear structures in mammograms. In H.-O. Peitgen, editor, *Proceedings of the 6th International Workshop on Digital Mammography (IWDM 2002)*, pages 285–291, Bremen, Germany, June 2002. Springer. 111

[139] L. C. C. Wai, M. Mellor, and M. Brady. A multi-resolution CLS detection algorithm for mammographic image analysis. In C. Barillot, D. R. Haynor, and P. Hellier, editors, *Lecture Notes in Computer Science, Proceedings of Medical Image Computing and Computer-Assisted Intervention (MICCAI 2004)*, pages 865–872, Berlin, Germany, 2004. Springer-Verlag. 112

[140] P. Miller and S. Astley. Detection of breast asymmetry using anatomical features. In R. S. Acharya and C. B. Goldgof, editors, *Biomedical Image Processing and Biomedical Visualization*, volume 1905 of *Proceedings of SPIE*, pages 433–442, San Jose, CA, February 1993. SPIE. 112

[141] P. Miller and S. Astley. Automated detection of breast asymmetry using anatomical features. In K. W. Bowyer and S. Astley, editors, *State of the Art in Digital Mammographic Image Analysis*, volume 9 of *Series in Machine Perception and Artificial Intelligence*, pages 247–261. World Scientific, River Edge, NJ, 1994. DOI: 10.1142/S0218001493000790 112

[142] T.-K. Lau and W. F. Bischof. Automated detection of breast tumors using the asymmetry approach. *Computers and Biomedical Research*, 24:273–295, 1991. DOI: 10.1016/0010-4809(91)90049-3 112

[143] R.M. Rangayyan, R.J. Ferrari, and A.F. Frère. Analysis of bilateral asymmetry in mammograms using directional, morphological, and density features. *Journal of Electronic Imaging*, 16(1):Article number 013003, January 2007. DOI: 10.1117/1.2712461 1, 112

[144] M. P. Sampat, G. J. Whitman, M. K. Markey, and A. C. Bovik. Evidence based detection of spiculated masses and architectural distortion. In J. M. Fitzpatrick and J. M. Reinhardt, editors, *Proceedings of SPIE Medical Imaging 2005: Image Processing*, volume 5747, pages 26–37, San Diego, CA, April 2005. 115

[145] Q. Guo, J. Shao, and V. Ruiz. Investigation of support vector machine for the detection of architectural distortion in mammographic images. *Journal of Physics: Conference Series*, 15:88–94, 2005. DOI: 10.1088/1742-6596/15/1/015 115

[146] G. D. Tourassi, D. M. Delong, and C. E. Floyd Jr. A study on the computerized fractal analysis of architectural distortion in screening mammograms. *Physics in Medicine and Biology*, 51(5):1299–1312, 2006. DOI: 10.1088/0031-9155/51/5/018 115

[147] T. Matsubara, T. Ichikawa, T. Hara, H. Fujita, S. Kasai, T. Endo, and T. Iwase. Automated detection methods for architectural distortions around skinline and within mammary gland on mammograms. In H. U. Lemke, M. W. Vannier, K. Inamura, A. G. Farman, K. Doi, and J. H. C. Reiber, editors, *International Congress Series: Proceedings of the 17th International Congress and Exhibition on Computer Assisted Radiology and Surgery*, pages 950–955, London, UK, June 2003. Elsevier. 115

[148] N. R. Mudigonda and R. M. Rangayyan. Texture flow-field analysis for the detection of architectural distortion in mammograms. In A. G. Ramakrishnan, editor, *Proceedings of Biovision*, pages 76–81, Bangalore, India, December 2001. 115

[149] N. Eltonsy, G. Tourassi, and A. Elmaghraby. Investigating performance of a morphology-based CAD scheme in detecting architectural distortion in screening mammograms. In H. U. Lemke, K. Inamura, K. Doi, M. W. Vannier, and A. G. Farman, editors, *Proceedings of the 20th International Congress and Exhibition on Computer Assisted Radiology and Surgery (CARS 2006)*, pages 336–338, Osaka, Japan, June 2006. Springer. 116

[150] F. J. Ayres and R. M. Rangayyan. Characterization of architectural distortion in mammograms. In *Proceedings of the 25th Annual International Conference of the IEEE Engineering in Medicine and Biology Society (CD-ROM)*, pages 886–889, Cancún, Mexico, September 2003. DOI: 10.1109/MEMB.2005.1384102 116

[151] F. J. Ayres and R. M. Rangayyan. Detection of architectural distortion in mammograms via analysis of phase portraits and curvilinear structures. In *Proceedings of the 3rd European Medical and Biological Engineering Conference, IFMBE European Conference on Biomedical Engineering (CD-ROM)*, volume 11, Prague, Czech Republic, November 2005. paper number 1873, six pages on CD-ROM. 116

[152] E. A. Sickles. Mammographic features of 300 consecutive nonpalpable breast cancers. *American Journal of Roentgenology*, 146(4):661–663, 1986. 61

[153] H. C. Burrell, D. M. Sibbering, A. R. M. Wilson, S. E. Pinder, A. J. Evans, L. J. Yeoman, C. W. Elston, I. O. Ellis, R. W. Blamey, and J. F. R. Robertson. Screening interval breast cancers: Mammographic features and prognostic factors. *Radiology*, 199(4):811–817, 1996. 61

[154] M. Sameti, J. Morgan-Parkes, R. K. Ward, and B. Palcic. Classifying image features in the last screening mammograms prior to detection of a malignant mass. In N. Karssemeijer, M. Thijssen, J. Hendriks, and L. van Erning, editors, *Proceedings of the 4th International*

Workshop on Digital Mammography, pages 127–134, Nijmegen, The Netherlands, June 1998. 116

[155] N. Petrick, H. P. Chan, B. Sahiner, M. A. Helvie, and S. Paquerault. Evaluation of an automated computer-aided diagnosis system for the detection of masses on prior mammograms. In *Proceedings of SPIE Volume 3979, Medical Imaging 2000: Image Processing*, pages 967–973, San Diego, CA, 2000. DOI: 10.1117/12.387600 117

[156] B. Zheng, W. F. Good, D. R. Armfield, C. Cohen, T. Hertzberg, J. H. Sumkin, and D. Gur. Performance change of mammographic CAD schemes optimized with most-recent and prior image databases. *Academic Radiology*, 10:283–288, 2003. DOI: 10.1016/S1076-6332(03)80102-2 117

[157] E. S. Burnside, E. A. Sickles, R. E. Sohlich, and K. E. Dee. Differential value of comparison with previous examinations in diagnostic versus screening mammography. *American Journal of Roentgenology*, 179:1173–1177, 2002. 118

[158] S. Ciatto, M. R. Del Turco, P. Burke, C. Visioli, E. Paci, and M Zappa. Comparison of standard and double reading and computer-aided detection (CAD) of interval cancers at prior negative screening mammograms: blind review. *British Journal of Cancer*, 89:1645–1649, 2003. DOI: 10.1038/sj.bjc.6601356 118

[159] J. J. James. The current status of digital mammography. *Clinical Radiology*, 59:1–10, 2004. DOI: 10.1016/j.crad.2003.08.011 118

[160] E. D. Pisano. Current status of full-field digital mammography. *Radiology*, 214:26–28, 2000. 118

[161] M. J. Yaffe. Development of full field digital mammography. In N. Karssemeijer, M. Thijssen, J. Hendriks, and L. van Erning, editors, *Digital Mammography*, pages 3–10, Nijmegen, The Netherlands, June 1998. Kluwer Academic Publishers. 118

[162] J. M. Lewin, C. J. D'Orsi, R. E. Hendrick, L. J. Moss, P. K. Isaacs, A. Karellas, and G. R. Cutter. Clinical comparison of full-field digital mammography and screen-film mammography for detection of breast cancer. *American Journal of Roentgenology*, 179:671–677, 2002. 118

[163] M. O. Honda, P. M. de Azevedo Marques, and J. A. H. Rodrigues. Content-based image retrieval in mammography: using texture features for correlation with BI-RADS categories. In H.-O. Peitgen, editor, *Proceedings of the 6th International Workshop on Digital Mammography: IWDM 2002*, pages 231–233, Bremen, Germany, January 2002. Springer. 118, 119

[164] T. Nakagawa, T. Hara, H. Fujita, T. Iwase, and T. Endo. Image retrieval system of mammographic masses by using local pattern matching technique. In H.-O. Peitgen, editor, *Proceedings of the 6th International Workshop on Digital Mammography: IWDM 2002*, pages 562–565, Bremen, Germany, January 2002. Springer. 118, 119

[165] S. K. Kinoshita, P. M. de Azevedo Marques, R. R. Pereira Jr, J. A. H. Rodrigues, and R. M. Rangayyan. Content-based retrieval of mammograms using visual feaures related to breast density patterns. *Journal of Digital Imaging*, 20(2):172 – 190, June 2007. DOI: 10.1007/s10278-007-9004-0 118

[166] X. Wang, M. R. Smith, and R. M. Rangayyan. Mammographic information analysis through association-rule mining. In *Proceedings of the IEEE Canadian Conference on Electrical and Computer Engineering*, pages 1495–1498, Niagara Falls, ON, May 2004. DOI: 10.1109/CCECE.2004.1349689 118

[167] H. Alto, R. M. Rangayyan, and J. E. L. Desautels. Content-based retrieval and analysis of mammographic masses. *Journal of Electronic Imaging*, 14(2):Article number 023016, pp. 1–17, 2005. DOI: 10.1117/1.1902996 119

[168] H. Alto, R. M. Rangayyan, R. B. Paranjape, J. E. L. Desautels, and H. Bryant. An indexed atlas of digital mammograms for computer-aided diagnosis of breast cancer. *Annales des Télécommunications*, 58(5-6):820–835, 2003. DOI: 10.1007/BF03001532 119

[169] H. Alto, R. M. Rangayyan, R. B. Paranjape, J. E. L. Desautels, and H. Bryant. An indexed atlas of digital mammograms for computer-aided diagnosis of breast cancer. In J. S. Suri and R. M. Rangayyan, editors, *Recent Advances in Breast Imaging, Mammography, and Computer-Aided Diagnosis of Breast Cancer*, pages 109–127. SPIE Press, Bellingham, WA, 2006. 119

[170] D. Guliato, E. V. de Melo, R. S. Bôaventura, and R. M. Rangayyan. AMDI - indexed atlas of digital mammograms that integrates case studies, e-learning, and research systems via the web. In J. S. Suri and R. M. Rangayyan, editors, *Recent Advances in Breast Imaging, Mammography, and Computer-aided Diagnosis of Breast Cancer*, pages 529–555. SPIE Press, Bellingham, WA, 2006. 119

[171] K. Doi. Diagnostic imaging over the last 50 years: research and development in medical imaging science and technology. *Physics in Medicine and Biology*, 51:R5–R27, June 2006. DOI: 10.1088/0031-9155/51/13/R02 99, 119

[172] K. Doi. Computer-aided diagnosis in medical imaging: historical review, current status and future potential. *Computerized Medical Imaging and Graphics*, 31:198–211, 2007. DOI: 10.1016/j.compmedimag.2007.02.002 99, 119

[173] R. M. Rangayyan, F. J. Ayres, and J. E. L. Desautels. A review of computer-aided diagnosis of breast cancer: toward the detection of subtle signs. *Journal of the Franklin Institute*, 344:312–348, 2006. DOI: 10.1016/j.jfranklin.2006.09.003 99, 119

[174] R2 Technology website, http://www.r2tech.com/, accessed on March 6, 2005. 99, 120

[175] S. M. Astley and F. J. Gilbert. Computer-aided detection in mammography. *Clinical Radiology*, 59:390–399, 2004. DOI: 10.1016/j.crad.2003.11.017 120

[176] J. J. Fenton, S. H. Taplin, P. A. Carney, L. Abraham, E. A. Sickles, C. D'Orsi, E. A. Berns, G. Cutter, W. E. Barlow R. E. Hendrick, and J. G. Elmore. Influence of computer-aided detection on performance of screening mammography. *The New England Journal of Medicine*, 356(14):1399–1409, April 2007. DOI: 10.1056/NEJMoa066099 120

[177] S. Ciatto, M. R. Del Turco, G. Risso, S. Catarzi, R. Bonaldi, V. Viterbo, P. Gnutti, B. Guglielmoni, L. Pinelli, A. Pandiscia, F. Navarra, A. Lauria, R. Palmiero, and P. L. Indovina. Comparison of standard reading and computer aided detection (CAD) on a national proficiency test of screening mammography. *European Journal of Radiology*, 45:135–138, 2003. DOI: 10.1016/S0720-048X(02)00011-6 120

[178] T. W. Freer and M. J. Ulissey. Screening mammography with computer-aided detection: Prospective study of 12,860 patients in a community breast center. *Radiology*, 220:781–786, 2001. DOI: 10.1148/radiol.2203001282 121

[179] L. J. W. Burhenne, S. A. Wood, C. J. D'Orsi, S. A. Feig, D. B. Kopans, L. F. O'Shaughnessy, E. A. Sickles, L. Tabar, C. J. Vyborny, and R. A. Castellino. Potential contribution of computer-aided detection to the sensitivity of screening mammography. *Radiology*, 215(2):554–562, 2000. 121

[180] W. P. Evans, L. J. W. Burhenne, L. Laurie, K. F. O'Shaughnessy, and R. A. Castellino. Invasive lobular carcinoma of the breast: Mammographic characteristics and computer-aided detection. *Radiology*, 225(1):182–189, 2002. DOI: 10.1148/radiol.2251011029 121

[181] R. L. Birdwell, D. M. Ikeda, K. F. O'Shaughnessy, and E. A. Sickles. Mammographic characteristics of 115 missed cancers later detected with screening mammography and the potential utility of computer-aided detection. *Radiology*, 219(1):192–202, 2001. 121

[182] J. A. Baker, E. L. Rosen, J. Y. Lo, E. I. Gimenez, R. Walsh, and M. S. Soo. Computer-aided detection (CAD) in screening mammography: Sensitivity of commercial CAD systems for detecting architectural distortion. *American Journal of Roentgenology*, 181:1083–1088, 2003. 121

[183] M. J. M. Broeders, N. C. Onland-Moret, H. J. T. M. Rijken, J. H. C. L. Hendriks, A. L. M. Verbeek, and R. Holland. Use of previous screening mammograms to identify features indicating cases that would have a possible gain in prognosis following earlier detection. *European Journal of Cancer*, 39(12):1770–1775, 1993. DOI: 10.1016/S0959-8049(03)00311-3 121

[184] R. C. Gonzalez and R. E. Woods, *Digital Imaging Processing*, Prentice-Hall, 2002. 100

[185] C. Serrano, B. Acha, R. M. Rangyyan, J. E. L. Desautels, Detection of microcalcifications in mammograms using error or prediction and statistical measures, *Journal of Electronic Imaging*, 18(1):1-10, 2009. DOI: 10.1117/1.3099710 105

[186] R. M. Rangayyan, S. Prajana, F. J. Ayres, and J. E. L. Desautels, Detection of Architectural Distortion in Mammograms Acquired Prior to the Detection of Breast Cancer using Gabor Filters, Phase Portraits, Fractal Dimension, and Texture Analysis, *International Journal of Computer Assisted Radiology and Surgery*, 2(6):347–361, 2008. DOI: 10.1007/s11548-007-0143-z 117

[187] S. Banik, R. M. Rangayyan, and J. E. L. Desautels, Detection of Architectural Distortion in Prior Mammograms of Interval-cancer Cases with Neural Networks, *Proceedings of the 31st Annual International Conference of the IEEE Engineering in Medicine and Biology Society*, 6667–6670, 2009. DOI: 10.1109/IEMBS.2009.5334517 117

[188] R. M. Rangayyan, S. Banik, S. Prajna, and J. E. L. Desautels, Detection of architectural distortion in prior mammograms of interval-cancer cases, 4(1):S171–S173, 2009. DOI: 10.1109/IEMBS.2009.5334517 117

[189] S. Banik, R. M. Rangayyan, J. E. L. Desautels, Detection of architectural distortion in prior mammograms or interval cancer using Laws' texture energy measures, *Proceedings of the 24th International Congress and Exhibition on Computer Assisted Radiology and Surgery*, Spring, 2010. 117

[190] P. M. de Azevedo-Marques, N. A. Rosa, A. J. M. Traina, C. Traina Junior, S. K. Kinoshita, and R. M. Rangayyan, Reducing and Semantic Gap in Content-based Image Retrieval in Mammography with Relevance Feedback in Inclusion of Expert Knowledge, *International Journal of Computer Assistes Radiology and Surgery*, 3(1):123–130, 2008. DOI: 10.1007/s11548-008-0154-4 119

[191] S. Banik, R. M. Rangayyan, and J. E. L. Desautels, "Detection of Architectural Distortion in Prior Mammograms Using Fractal Analysis and Angular Spread of Power", *Proceedings of SPIE Medical Imaging 2010: Computer-Aided Diagnosis*, San Diego, CA, February 2010. Ed. edited by N. Karssemeijer, R. M. Summers, Vol. 7624, 762408:1-9. 117

[192] R. M. Rangayyan, S. Banik, and J. E. L. Desautels, "Computer-aided Detection of Architectural Distortion in Prior Mammograms of Interval Cancer", *Journal of Digital Imaging*, 23(5):611-631, October 2010. DOI: 10.1007/s10278-009-9257-x 117

[193] D. Guliato, R. S. Bôaventura, M. Maia, R. M. Rangayyan, M. S. Simedo, and T.A.A. Macedo, "INDIAM - An e-Learning System for the Interpretation of Mammograms", Journal of Digital Imaging, 22(4):405–420, August 2009. DOI: 10.1007/s10278-008-9111-6 119

Authors' Biographies

FÁBIO J. AYRES

Fábio J. Ayres obtained his B.Sc. in Electrical Engineering at the University of São Paulo, São Paulo, Brazil, in 1997 and his M.Sc. degree in Electrical Engineering at the same university in 2001. He obtained his Ph.D. in 2007 at the University of Calgary, Calgary, Canada. His research interests are image-based computer-aided diagnosis, computer vision, and surgical simulation.

RANGARAJ M. RANGAYYAN

Rangaraj M. Rangayyan is a Professor with the Department of Electrical and Computer Engineering, and an Adjunct Professor of Surgery and Radiology, at the University of Calgary, Calgary, Alberta, Canada. He received the Bachelor of Engineering degree in Electronics and Communication in 1976 from the University of Mysore at the People's Education Society College of Engineering, Mandya, Karnataka, India, and the Ph.D. degree in Electrical Engineering from the Indian Institute of Science, Bangalore, Karnataka, India, in 1980. His research interests are in the areas of digital signal and image processing, biomedical signal analysis, biomedical image analysis, and computer-aided diagnosis. He has published more than 140 papers in journals and 220 papers in proceedings of conferences. His research productivity was recognized with the 1997 and 2001 Research Excellence Awards of the Department of Electrical and Computer Engineering, the 1997 Research Award of the Faculty of Engineering, and by appointment as a "University Professor" in 2003, at the University of Calgary. He is the author of two textbooks: *Biomedical Signal Analysis* (IEEE/ Wiley, 2002) and *Biomedical Image Analysis* (CRC, 2005); he has coauthored and coedited several other books. He was recognized by the IEEE with the award of the Third Millennium Medal in 2000, and was elected as a Fellow of the IEEE in 2001, Fellow of the Engineering Institute of Canada in 2002, Fellow of the American Institute for Medical and Biological Engineering in 2003, Fellow of SPIE: the International Society for Optical Engineering in 2003, Fellow of the Society for Imaging Informatics in Medicine in 2007, Fellow of the Canadian Medical and Biological Engineering Society in 2007, and Fellow of the Canadian Academy of Engineering in 2009. He has been awarded the Killam Resident Fellowship thrice (1998, 2002, and 2007) in support of his book-writing projects.

J.E. LEO DESAUTELS

J.E. Leo Desautels obtained his M.D. from the University of Ottawa in 1955, and completed post-graduate training in radiology at the Henry Ford Hospital, Detroit, MI. He was a Staff Radiologist at the Foothills Hospital and a Clinical Professor with the Faculty of Medicine, the University of Calgary, Calgary, AB, Canada, from 1970 to 1994. He served as a Reference Radiologist to the Alberta Program for the Early Detection of Breast Cancer until 2007. He is an Adjunct Professor of Electrical and Computer Engineering at the University of Calgary. He is interested in computer applications in mammography.

Index